A series of student texts in

CONTEMPORARY BIOLOGY

General Editors:
Professor E. J. W. Barrington, F.R.S.
Professor Arthur J. Willis

Theories of
Differentiation

Max Hamburgh
M.A. (Yale), Ph.D. (Columbia)

Department of Biology, City College of New York,
and
Department of Anatomy, Albert Einstein College of Medicine

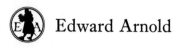 Edward Arnold

First published 1971
by Edward Arnold (Publishers) Ltd.,
41 Maddox Street,
London, W1R 0AN

Boards Edition ISBN: 0 7131 2320 6
Paper Edition   ISBN: 0 7131 2321 4

Printed in Great Britain by
William Clowes & Sons, Limited, London, Beccles and Colchester

Preface

The last few years have seen the appearance of books on development in ever increasing numbers. The full extent of this interest in the subject and the attempts of the book market to satisfy it can be appreciated by a quick glance at the list of general references. In the light of this, the presentation of another elementary text book requires some justification.

Like the Japanese fable of *Rashamon*, where three different eyewitnesses each gave a very different account of the same event, so the story of development, like any big story, can only be fully comprehended if many different observers each tell their own version. My approach to development may seem a little unusual to some, but it has been planned with the deliberate intention of presenting only those topics of developmental biology which have been fitted into well-defined conceptual models, and for which major generalizations have been advanced. For this reason, the morphology of development, already covered very fully in recent texts, has been omitted. With the cracking of the genetic code achieved, the problem of the nature and mechanism of cell differentiation has become a major question of modern developmental biology, and it is upon this question that I have chosen to concentrate. The search for the answers to it is not a recent undertaking, but has a considerable background and history.

Development can be studied (and has been) essentially in two ways. One approach is that of the silent observer and recorder, who puts himself and his material into an intellectual isolation booth and follows all the changes and transformations of the fertilized egg to the miniature adult. For reasons of convenience he may, rather than watching the unfolding of events continuously, decide to stop the process at intervals, fix his material, section and stain, and then literally reconstruct the story from his slide collection. The knowledge gained from this procedure— the so-called 'descriptive embryology'— is invaluable and constitutes the critical mass of information which the biologist must acquire before he can even contemplate the analysis of underlying principles.

The other approach, the approach of 'experimental embryology' utilized in this book, is designed to identify mechanisms and to analyse the control of developmental events and processes. For this it is not enough merely to observe and record; the investigator must also be a detective. Like a sleuth, the scientist must subject his material (in this case differentiating cell populations or developing organisms) to some well-defined, often outlandish, situations, in order to test reactions and responses and so confirm or disprove his suspicions.

The choice of experimental material is mainly a matter of convenience. Just as neurophysiologists use the squid because its gigantic axons (1 mm in diameter) lend themselves to having electrodes inserted in them more than the nervous systems of other beings (e.g. *Drosophila*), so embryologists prefer sea urchin, salamander, frog and chick eggs, in that order of preference, because they are abundant, accessible, and, unlike mammals, develop in plain view instead of inside a maternal uterus, though recently, the application of tissue culture to studies of development has brought the mammalian embryo also into the range of the experimental biologist. Fortunately, it does not really matter very much which organism one ultimately selects for the study of biological processes.

The study of morphology teaches us the diversity of living forms; so, too, biochemistry and molecular biology must increasingly impress us with the unity of life. Nature, at least as it relates to the operation of life processes, is rather conservative. Though the genetic code was revealed mainly by the study of viruses and *E. coli*, there are good reasons for believing it to apply universally for all organisms. The mechanism of hereditary transmission was discovered by breeding experiments with *Drosophila*, but the information gained applies equally to plants and primates. The biochemistry of metabolism, on which so much of our modern drug therapy is based, was worked out on yeast. There is every reason to believe that the principles of development that are being revealed by manipulation with sea urchin, salamander, frog and chick embryos, apply with minor variations equally well to all beasts, including man.

The present book discusses and evaluates a great many theories and hypotheses. It is quite conceivable that within a short time many of them will be obsolete, and that almost all of them will be proven wholly or partially wrong. However, a theory should not be judged on whether it expresses total and unchanging truth. The goodness-of-fit of a theory should be judged on three counts: (1) How well does it explain the observations available at the time, (2) Can the explanation be turned around to predict events, and (most important of all) (3) Does the theory or hypothesis advanced lend itself to the formulation of new tests and subsequent experiments? Many hypotheses will satisfy one or other of the above criteria; few will satisfy all. The one which I consider most important, and which I have favoured in the selection of material in this book, is the last, for any theory that stimulates new experiments is worthy of attention. At the risk of sounding pompous, I hope most of all that I will succeed in infecting the student of biology with some of the excitement and ferment that goes on in the field. Every science has its periods of breakthrough and its periods of consolidation. As physics had its great period in the 1930s and 1940s, so the present is the 'Age of Experimental Biology', with particular emphasis on developmental biology. It is my belief that students of development will eventually provide the keys to unlock not only the riddle of growth and differentiation but also of many other mysteries of life.

New York,   M.H.
1971

Acknowledgements

This book is dedicated to many people, too numerous to name, who have been good to me. It is dedicated first and foremost to my wife, from whom I stole the hours that were spent in writing it. It is dedicated also to the memory of my father and mother, who very early in life implanted the illusions in me that gave me an overdose of self-confidence. It is dedicated to my brother, who at a crucial moment in my life pushed me into a career in science. Grateful acknowledgement is also made to Miss Cassandra Kirk and Mrs. Rita Berkowitz for their help in the preparation of this manuscript. Last but not least, I dedicate this book to the memory of my teacher and mentor, the late Ernst Albert Scharrer, who remains my ideal of what a true scientist and humanist should be. Grateful acknowledgement is also made to Dr. Arthur and Dr. Laura Colwin, Dr. Marie Di Berardino, Dr. Philip Grant and Dr. John Saunders for their critical reading of this manuscript.

M.H.

Table of Contents

I

Interaction of Egg and Sperm

INTRODUCTION

Statement of problem

In his classic work *The Cell in Development and Inheritance*, E. B. Wilson wrote that the 'essential phenomenon of fertilization is the union of a sperm nucleus of paternal origin with an egg nucleus of maternal origin, to form the primary nucleus of the embryo.' However, that cannot be the whole story. The sperm can be dispensed with and development of the eggs of many species can still be initiated by a variety of other agents.

Frog eggs will respond to the pricking of a needle dipped in blood and start the developmental sequence that will carry them all the way to the tadpole stage.

Sea urchin eggs can be activated by exposure to acids, hyper- and hypotonic solution. Mammalian eggs will cleave and form a blastodisk in response to temperature shock.

All of these treatments that mimic the job normally performed by sperm and succeed in parthenogenetically activating the egg have this much in common, namely, that apparently an unspecific stimulus such as temperature shock, pH shock and hyper- and hypotonicity, or just a needle prick can trigger off a series of reactions in the egg which allows development and differentiation to proceed.

In the economy of nature, the trigger is usually, but not exclusively, pulled by the carrier of the paternal set of chromosomes, i.e., the sperm (rotifer eggs and drones developing from unfertilized ova are exceptions).

For some unknown reason, the spermatozoon does the job better than either shock treatment or needle pricking, as witnessed by the high degree of success obtained by the sperm as compared to the low degree of activation obtained with artificial agents.

All of these considerations imply, therefore, that the egg itself contains all the essentials for development. It must possess both the programme and the equipment necessary for making an embryo. The egg, therefore, far from being an 'undifferentiated' cell, is perhaps the most highly specialized cell of any organism. Yet, every schoolboy also knows that the mature ovum is a cell under a sentence of death that can only be commuted by the act of fertilization. The human egg must be fertilized between 24–48 hours after it has been shed into the Fallopian tube. Amphibian eggs that are retained too long within the body cavity of the female are poor prospects for fertilization and once they are shed into the water must be penetrated by sperm immediately or they will rapidly decline. In a way the egg resembles a piece of highly complicated machinery ready to spring into action once the switch is turned on, but one that will disassemble if the switch fails.

The puzzling question facing the embryologist as he starts the analysis of development is just precisely what does the sperm do to the egg to start it on its predetermined journey.

Structure of egg and sperm cells

The surface of a ripe egg is seldom naked, but it is usually surrounded by a noncellular barrier like the *jelly coat*, or the *vitelline membrane* and often there is an additional cellular barrier present like the follicle cells that surround mammalian eggs, as the *corona radiata*. (See Figs. 1.1., 1.2.)

In the echinoderms and frogs, the barrier takes the form of a gelatinous coat around the unfertilized egg which the sperm must penetrate to reach the egg surface. The salmon egg is surrounded by a tough outer membrane, the *chorion*, which is impermeable to the spermatozoa except at one point, the micropyle. In mammals, the *zona pellucida* and the corona radiata present two barriers that must be penetrated before successful fertilization can occur. These outer barriers are the features that lend most distinction to the egg. Otherwise, the morphology of the ovum differs little from that of other cells and contains nothing to reveal the extraordinary potential of this most unusual cell.

The other player is the sperm. The appearance of the sperm is the result of a series of transformations of the haploid spermatid cell. During the process called *spermiogenesis*, the nuclear material of the spermatid is compacted to form the bulk of the head of the sperm, the cyto-

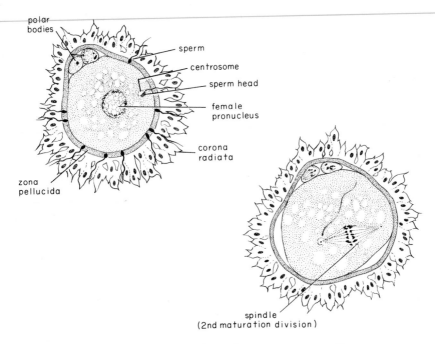

Fig. 1.1 Diagrammatic representation of a mature mammalian ovum surrounded by two barriers, the zona pellucida and the corona radiata. On the right is shown the same ovum after sperm entrance.

plasm is reduced in bulk and turns into an envelope which forms the *middle piece* and *tail*, and the cytoplasmic organelle known as the Golgi material is placed like a cap over the nuclear head of the sperm to form the *acrosome*.

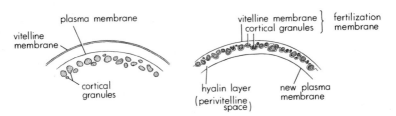

Fig. 1.2 Diagrammatic representation of an unfertilized sea urchin egg (left) and a fertilized sea urchin egg (right). A new coat of the fertilization membrane has formed. The latter is the product of fusion of the vitelline coat with the plasma membrane proper plus the cortical granules that have opened up.

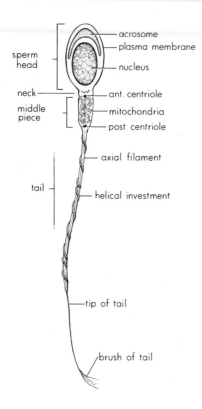

sperm head
— acrosome
— plasma membrane
— nucleus

neck —— ant. centriole

middle piece
— mitochondria
— post centriole

— axial filament

tail
— helical investment

—tip of tail

brush of tail

Fig. 1.3 Diagrammatic representation of human sperm. Total
length 50 μm. The neck, middle
piece and tail do not consist of a
single flagellum but are composed
of several strands.

The statistical probability of the event of fertilization

One of the problems nature had to solve was to convert so improbable
an event as fertilization into one which will occur with great certainty.
The problem is treated statistically and solved by increasing the number
of trials. The number of reproductive cells shed by most organisms is
tremendous, but though both egg and sperm are liberated in great
numbers, only a fraction of the eggs that are released eventually become
fertilized, and of those that become fertilized only a fraction survive the
rigours of development to emerge as mature adults. Considerations like
these led Darwin to formulate the Theory of Natural Selection. He
pondered what would happen if all the eggs shed by all the known species
of frog were to be fertilized, and if all of these were to survive to become
adults. He concluded that this planet long ago would have become too
small for any genus other than *Rana*.

In mammals the possibility of fertilization is enhanced by a mechanism that assures sperm deposition within the female genital tract, but the obstacles to successful fertilization are still so manifold that the number of sperm deposited far exceeds the number that eventually will reach the ovum. The average human adult male ejaculate contains approximately 60 million sperm per millilitre of seminal fluid. Assuming that the average male ejaculate is composed of 3 millilitres of seminal fluid, approximately 180 million to 200 million sperm are introduced into the female genital tract in order to assure penetration of one egg by one sperm. Gamow calculated the length of the trip that the human sperm has to complete from the point of deposit in the female tract until it reaches its target, the egg, and concluded that the journey of the sperm through the female tract on its way to the ovum is comparable to a trip by foot from New York to California. Chances are that of 100 people starting out on such a journey, most would drop out quite soon, others somewhat later, and only a very small number would eventually reach their destination.

SPERM-EGG INTERACTION

Lillie's discovery of fertilizin

Fertilization implies the penetration of one cell by another. To biologists, this is a most unlikely event, for even large molecules do not ordinarily gain entrance into the interior of a cell. A cell like the egg guarded not only by its own plasma membrane but surrounded by barriers such as the vitelline membrane, jelly coat and/or zona pellucida must be attacked at length before an invader like the sperm can gain entrance. Unless one wants to assume that pure chance collision between two cells can ensure the conditions necessary to enable the sperm to attack and pierce the egg barriers, one is in need of an hypothesis that can explain the mechanism by which egg-sperm attachment is brought about. Presumably, the sperm must attach to the egg and remain attached to it long enough to carry out the time-requiring reaction that leads to the destruction of the egg barriers, prior to its entry into the ovum.

The first information regarding the nature of sperm-egg attachment came in 1912 with the experiments conducted by F. R. Lillie. Lillie[184] discovered that sperm of sea urchins that were placed into a container of sea water in which eggs had previously been bathed, tended to agglutinate very rapidly in the egg water environment.

A biologist invariably associates *agglutination* or clumping of cells with antigen and antibody immune reactions. Lillie[184–187] suggested that perhaps the jelly coat and the surface of the sea urchin egg are covered with *receptor molecules* whose spatial configuration is

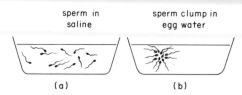

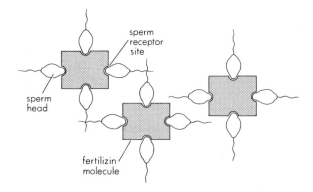

Fig. 1.4 Diagrammatic representation of Lillie's experiment. **(a)** Sea urchin sperm in a vial of sea water are freely motile. **(b)** When placed into a vial of egg water they immediately begin to clump.

complementary to the surface molecules covering the sperm, and he proposed the names *fertilizin* and *antifertilizin* respectively for these two hypothetical substances. He reasoned that fertilizin molecules diffusing into the sea water might then bind large numbers of sperm at their receptor sites, thus giving the appearance of sperm clumping that was first observed in his initial experiment.

Fig. 1.5 Diagrammatic representation of a multivalent fertilizin molecule having four active sites. The spatial configuration of the active sites are complementary to the configuration of the antifertilizins of the sperm. The lock and key arrangement permitting several sperm to fit into the receptor sites of one fertilizin molecule gives rise to the clumping effect.

Fertilizin and the antigen-antibody model

In keeping with the modern immunological theory, agglutination is considered the result of chemical interaction between two complementary substances whose spatial configurations are like a lock and key arrangement. Agglutination of sperm by egg fertilizin implies the existence of a complementary receptor substance, antifertilizin, on the sperm. Such a substance should have biological properties similar to that of the fertilizin of the eggs. It should occur in an appreciable amount on the sperm surface, in solution it should neutralize fertilizin, precipitate egg jellies, and agglutinate eggs. Such a substance located on the sperm surface was indeed demonstrated by Lillie[184] and some of the chemical properties of fertilizin and antifertilizin have since been identified. (For a review, see Metz,[193, 194] Tyler,[261, 262] and Rothschild.[221])

The spermatozoa are agglutinated by their heads, indicating that the receptor molecules of the sperm are located at the surface of the head. Tyler has calculated that the head of a sea urchin egg can easily accommodate 10^5 fertilizin molecules.

The cluster of agglutinated sperm breaks up after a while, and such disagglutinated sperm, though still mobile, can not be reagglutinated again by fertilizin of fresh jelly coat solution; their fertilizability is also greatly reduced. Possibly the reversal of the agglutination is the result of breakdown of jelly coat molecules so that only one active group or **uni- valent** molecule remains attached to the sperm. Failure of the sperm to reagglutinate is believed to result from combination of univalent fertilizin with the antifertilizin of the sperm surface (Tyler[263]).

It is pertinent to mention that jelly coat substance when treated with proteolytic enzymes such as trypsin and chymotrypsin loses agglutinating power. Most probably proteinases split the jelly coat macromolecule into smaller ones (at the same site where they are split during reversal). Each of the fractions obtained from proteolytic digestion still carries one combining group, and, therefore, can still attach to the surface of the sperm. However, only the binding of many sperms to a multivalent molecule results in the agglutination reaction, which is visible to the naked eye as clumping of cells. (See Fig. 1.5.)

Chemistry of fertilizin and antifertilizin

Chemically, fertilizin is a glyco-protein. The monosaccharides found are glucose, fructose, and galactose. Both the amino acids and the monosaccharides vary from species to species so that it is more correct to speak of fertilizins. The fertilizin molecule is a relatively large one (M. Wt. 300 000).

Tyler[262] analysed the chemistry of fertilizin using such tests as dialysis, electrophoresis and enzyme degradation and concluded that the amino acids are scattered or clustered as peptides at certain strategic points between the monosaccharide residues. There is a high species specificity of the agglutinating reaction of fertilizin with very little cross reaction. Fertilizin of species A will agglutinate only sperm of species A. This high specificity is not reflected in any way in different compositions of either sugars or amino acids, but it is more likely a consequence of differences in the molecular configurations. For a review of the chemistry of fertilizin, see Metz,[192–194] Monroy,[197] Runnstrom, Hagstrom and Perlman,[227] and Tyler.[263]

The available information on the chemistry of antifertilizin is more scanty. Antifertilizin is a heat-stable acidic protein, and a much smaller molecule than fertilizin (M. Wt. 10 000). Its presence is routinely demonstrated by the ability of the extracted substance to agglutinate eggs in sea water or to neutralize the agglutination of sperm by egg water. Both effects are presumably due to the competition for fertilizin binding sites between sperm and extracts of antifertilizin (Metz,[193] Tyler[261]).

Site and sources of fertilizin

Jelly coat fertilizin

Lillie originally believed that fertilizin is continually emitted from mature sea urchin eggs. The main source of it is probably not the surface of the egg itself but rather the jelly coat or the vitelline membrane. It is generally assumed that the jelly coat consists probably entirely of fertilizin because, after the coat of sea urchin eggs has been dissolved in acidified sea water, the egg is depleted of practically all of its fertilizin. There is some argument as to the source of the jelly coat itself. The weight of evidence seems to favor the view that the follicles of the ovary are the source, because fertilizin can be obtained from spent ovaries of sea urchins that lack eggs.

The question of the source of fertilizin has been examined in two forms other than sea urchins, the gastropod *Megathuria* and the annelid *Nereis*. The unfertilized *Nereis* egg has no gelatinous coat, but upon activation, a thick jelly is extruded. Solutions of this jelly, which can be obtained by treating eggs with alkaline sea water, will agglutinate sperm effectively, suggesting that this jelly material may contain a high titre of fertilizin. Egg water obtained from unfertilized *Nereis* eggs contains some fertilizin also, but its sperm agglutination power is low. On the basis of these observations Tyler suggested that in *Nereis* some jelly material may always diffuse into the sea water but that fertilizin in *Nereis* is probably provided mainly by the egg cell itself rather than the follicle cells of the ovaries.

Cytofertilizin

There is good evidence that fertilizin is a component, not only of the egg barriers like the jelly coat but of the plasma membrane of some eggs also. Eggs deprived experimentally of their jelly coat, or eggs of species that normally lack a jelly coat, can be demonstrated with suitable means to possess fertilizin. Motomura[202] was able to extract sperm-agglutinins from the unfertilized jellyless egg of the species *Strongylocentrotus pulcherimus*. This fertilizin, for which the plasma membrane is suspected to be the source, is sometimes referred to as **cytofertilizin** in contradistinction to **jelly coat fertilizin**. The former is probably the real receptor substance of eggs for spermatozoa. Further evidence that fertilizin may be a component of the plasma membrane as well as of the jelly coat comes from experiments reported by Tyler and Brookbank.[268] They prepared rabbit antiserum against fertilizin and noted that such antisera can block division of fertilized sea urchin eggs, whereas antibodies directed against the interior of the egg are ineffective in blocking cell division. The antibodies evidently do not manage to get inside the sea urchin egg, but may combine with their antigens or fertilizin at the surface.

Biological significance of the fertilizin-antifertilizin reaction

Goodness of fit of the theory of fertilizin-mediated egg-sperm attachment

The linking of the fertilizin with antifertilizin, molecules that are presumably present on the surface of egg and spermatozoa respectively, could establish the initial bond necessary to give the sperm the time and chance to attack the egg barriers. The proof of the theory will depend on the goodness of fit of the evidence that the site and location of fertilizin molecules on the surface of the egg is such as to enable a prolonged link-up. If one is convinced, as some are, that fertilizin in sea urchin eggs is located only and exclusively in the jelly coat, from where it is extruded into the sea water, one will not find such a theory convincing. The most one will be able to agree to, is, probably, that the fertilizin diffusing into the environment may at best perform a kind of mopping-up operation and help to get rid of excess sperm that accumulate in the neighbourhood of the egg. In this way, fertilizin may serve to prevent polyspermy. If on the other hand, one is convinced that fertilizin is located also on the plasma membrane of the egg then the interaction between fertilizin and antifertilizin may provide the key mechanism by which sperm-egg attachment is assured.

The biological significance of the fertilizin-antifertilizin reaction is usually judged along three lines of evidence.

1 How universal is the presence of fertilizin in eggs of different animal species?

2 What effect, if any, does removal or addition of fertilizin exert on the fertilizability of eggs?

3 What effect, if any, does fertilizin exert on the fertilizability of sperm?

Distribution of fertilizin in the plant and animal kingdoms

With respect to occurrence and distribution of fertilizin, it can be stated that fertilizin has been found in four phyla, the molluscs, echinoderms, annelids, and chordates, and that the list is continuously being extended. The agglutination of sperm by egg water, however, must not be taken as the only criterion for the presence of fertilizin. In those species in which the fertilizin molecule exists in univalent forms, combinations of sperm with fertilizin will not be detectable by an agglutination reaction. Sperm of many species that fail to agglutinate after exposure to egg water, have been shown to lose their fertilizing power after such treatment (Tyler,[260] Metz[193]). Many species (for example, the starfish, and *Urechis*) that were formerly listed as lacking fertilizin because their egg extracts failed to agglutinate sperm, may merely represent cases in which this molecule is present in univalent form. Tyler[260] and Metz[192, 193] have provided considerable evidence for this view. They have extracted substances from eggs of non-agglutinating species, treated them with *adjuvants*, and subsequently succeeded in agglutinating sperm with such egg extracts. The effect of adjuvants is explained as being due to removal of some ions which may block surface groupings. Since addition of adjuvants to egg extracts can cause agglutination reactions in several species that normally fail to agglutinate, it is not unreasonable to postulate that fertilizin may be far more widely distributed in the animal kingdom than was first suspected. In any case, failure to demonstrate agglutination under ordinary conditions cannot be equated *per se* with lack of fertilizin. In some species, but not all, calcium is necessary to bring about agglutinating reactions. In other species, fertilizin-like substances may occur in insoluble form and be firmly bound to the egg surface. Sea urchin eggs, for example, appear to retain a bound layer of fertilizin at their surface long after the jelly coat has been removed by acid treatment and/or enzyme digestion. One may conclude, therefore, that fertilizin exists in several forms, that it requires a variety of methods to demonstrate it, and that it is probably widely distributed among gametes of all phyla.

Effects of removal and addition of fertilizin on the fertilizability of eggs

If fertilizin is essential for fertilization, then fertilizin-free eggs should fail to be fertilized and fertilizability should be restored upon addition of fertilizin. This test has been attempted by a number of workers, notably Tyler.[260] Fertilizin cannot be demonstrated by routine methods in acid-treated jellyless sea urchin eggs. Nevertheless, such eggs can be fertilized. This may mean that fertilizin is not essential for fertilization. An alternate explanation advanced by Tyler,[260] based on the observation that jellyless eggs do agglutinate when they are treated with antifertilizin from sperm, suggests that some fertilizin always remains bound to the egg surface and that it is this plasma-bound fertilizin (cytofertilizin) which plays the *major* role in fertilization. Tyler and Metz (quoted in Monroy[198]) applied crystalline trypsin to digest and remove jelly coat and vitelline membrane from *Arbacia* eggs and noted that such eggs can still be fertilized, but there is some loss of fertilizability which parallels the gradual loss of agglutinizability. Eggs depleted of much of their jelly coat fertilizin do show a much greater receptivity to sperm of foreign species, however. An unequivocal test for fertilizability of fertilizin-free eggs is still lacking, but from the data of Tyler and Metz it seems that fertilizability and agglutinizability of eggs decrease in parallel fashion with the length of enzyme digestion of egg fertilizin.

Effects of removal or blocking of antifertilizin from sperm

The counterpart of fertilizin removal from eggs is removal of anti-fertilizin from sperm. This was tried by Tyler and O'Melveny (quoted in Monroy[198]). By acid treatment they removed antifertilizin from sea urchin sperm and noted reduced fertilizing capacity.

Effect of fertilizin on fertilizing capacity of sperm

As already mentioned, sea urchin sperm treated with fertilizin will first agglutinate but after a while, a reversal of the agglutination reaction sets in (Tyler[260]). Blocking of the antifertilizin group of sperm with multivalent fertilizin or with antisera prepared against antifertilizin renders sperm incapable of fertilizing eggs any longer. Antisera produced against *Arbacia* sperm can be rendered univalent by papain digestion. Such univalent antisera when mixed with *Arbacia* sperm will not agglutinate the sperm, but render them incapable of fertilizing an egg. (For review see Metz.[194])

Possible function of fertilizin-antifertilizin interaction in the process of fertilization

Primary function

The wide distribution of these two molecules in sex cells in so wide a variety of animals, and the reduction of fertilizing potential by treatments that tend to remove these substances from gametes, convinced many biologists that to the interaction of these two antigens must be assigned a significant function in the reproductive process. Accordingly, the linking of egg fertilizin with the antifertilizin of the sperm's surface provides a model which could account for the establishment of the initial bond by which the two cells can attach and thus enable the sperm to attack the egg barriers.

Subsidiary functions of fertilizin

There are, however, other additional functions which these two surface molecules may serve. It has been stated before that the only way to assure fertilization is to increase the number of sperm that can reach the vicinity of the egg to a proportion high enough to raise the chance of egg-sperm interaction to a certainty. The excessive ratio of sperm to eggs has the inherent danger of promoting polyspermy. One way to prevent polyspermy is to thin the population of sperm that have succeeded in surrounding the egg to manageable proportions. Fertilizin leaking out from the egg surface may serve this function and act to mop up the excess sperm population, thus decreasing the chance of polyspermy. Another function which the fertilizin-antifertilizin reaction may fulfil is to provide a mechanism for prevention of interspecies fertilization. There is a high degree of cross-agglutination in those groups that can cross-fertilize, and a low degree of cross-agglutination between fertilizin and antifertilizin from species that do not normally interbreed in nature. While it is conceded that the wrong key can sometimes open a door, usually a lock that does not fit the key will tend to keep unwanted intruders out. A third function sometimes ascribed to the fertilizin reaction relates to the fact that sperm cells contain potent lytic agents. These lysing substances are presumably required to break down the barriers surrounding the egg. Holding large numbers of sperm to the egg surface while at the same time preventing wholesale invasion may supply the lytic enzymes in the quantities necessary to digest the barriers until one sperm eventually manages to slip in.

Sperm activation by fertilization

A fourth major property of fertilizin already observed by Lillie relates to the ability of this molecule to activate sperm. Sperm activation can easily be demonstrated by adding jelly coat solutions to a population of immotile sperm. Upon addition of jelly coat solution, aged, immotile spermatozoa revive and respond with an immediate outburst of motility, that will last for a much longer period than that of untreated sperm. Aged spermatozoa respond to treatment with jelly coat solution with quick increases in O_2 consumption. Possibly, fertilizin acts as a chelating agent removing inhibitors or certain toxic ions from the sea water.

The question has been raised whether the two properties, i.e. the ability to agglutinate and the capacity to activate sperm, reside both in the same or in different groups of the fertilizin molecule. Electrophoretic and ultracentrifugation analyses indicate that the jelly coat substance is homogenous. On the other hand, the sperm agglutinating power of fertilizin is lost upon heating, while the sperm activation effect is not affected by heat treatment. The activation factor is dialysable but the agglutination factor is not. Evidence has been presented that the substance responsible for raising sperm respiration, i.e. the activating factor, is a diffuse substance located on the surface of the egg itself. (For review see Metz[193] and Mintz.[195]) Although this substance can be extracted from jelly coat preparations, great quantities of it appear also in sea water in which jellyless *Arbacia* eggs have been standing. The distinction between activating molecules and receptor molecules strongly suggests the existence of two different types of fertilizin, which differ subtly in chemical configuration and also with respect to their source and origin. It is possible that both molecules are present in different proportions in the jelly coat. The activating molecule may be a major constituent of the egg surface whereas the greater proportion of the agglutinating molecule may be located on the vitelline membrane. The distinction between *activator* and *receptor* fertilizin, and the possibility of distinct location of these molecules, are also attested to by experiments in which antibodies have been produced against these two substances. Several different antigens have been identified in the egg surface of sea urchins. One of these belongs to the jelly coat, while another, the A antigen, which is heat-stable and is not attacked by trypsin, has its seat on the egg surface. This antigen may be identical with the activating molecule, for sperm activation is not destroyed by trypsin. Another antigen, the F antigen, is also found at the surface of the egg. Treatment of eggs with antibodies prepared against it will decrease the fertilizability of eggs.

The question of the significance of the fertilizin-antifertilizin reaction and its role in fertilization remains a source of controversy. Metz[194] in his excellent review, expressed the following judgment:

'Interaction of sperm and egg surfaces and the role of surface substances in fertilization have been investigated most thoroughly in sea urchins and amphibians among metazoans. The thinking on this subject has been dominated by the fertilizin-antifertilizin concept for over 50 years. Fertilizin is the major component of the sea urchin egg jelly. In solution it specifically agglutinates the sperm by combining with antifertilizin, a molecule located on the surface of the sperm. This specific isoagglutination has been demonstrated in some, but not all, species tested in several phyla. Several methods have revealed the same or comparable systems in other groups. Accordingly, it appears that a fertilizin-antifertilizin system is of wide, if not universal, occurrence. Evidence from a variety of experimental approaches indicates that a fertilizin-antifertilizin interaction is an important and probably essential step in fertilization. The fertilizin-antifertilizin interaction probably initiates the acrosomal reaction [See Chapter 2 p. 19], functions in attachment and fusion of

the gametes, and contributes a primary specificity factor in fertilization. In addition, surface antigens of the sperm, and probably the
egg as well, have some essential, but as yet unknown role in [sea
urchin] fertilization.'

THE POLYSPERMY-PREVENTING REACTION

The raising of the fertilization membrane in echinoderm eggs

One of the most important physiological expressions of the changes
that the egg surface undergoes at fertilization is the reaction that makes it
immediately unresponsive to additional spermatozoa after the first sperm
has made contact, thus preventing entry of more than one sperm into the
egg. Several mechanisms have been proposed to account for the prevention of **polyspermy**. It has been demonstrated, as pointed out above,
that antifertilizin is not only a constituent of sperm but of the egg also.
One function suggested for egg antifertilizin is that at fertilization the egg
antifertilizin may bind with its own surface fertilizin. After completion
of this process, additional spermatozoa can no longer attach to the egg
surface for lack of available binding sites. More than one mechanism
probably operates to prevent polyspermy. In order to guard against the
danger of excess sperm invasion, nature has, so to speak, written out a
multiple insurance policy by evolving a series of **polyspermy-preventing
reactions**. In the sea urchin egg, the first defense against polyspermy
probably resides in the jelly coat itself whose fertilizin might act like a
sieve on the approaching spermatozoa. The agglutination of excess
spermatozoa by fertilizin that leaks out continuously from this jelly coat
into the medium can thus reduce the number of sperm capable of
attacking the egg.

The major mechanism preventing polyspermy must be a split second
change in the barrier surrounding the egg, thus rendering it impermeable
to additional sperm immediately after the first sperm has entered. In
echinoderms, this change in accessibility is signalled by the raising of the
'fertilization membrane'. At fertilization, the vitelline membrane of
echinoderm eggs lifts off from the egg surface and fuses with the plasma
membrane of the ovum. In this process a new membrane is formed,
which is really a product of the fusion of vitelline and plasma membrane
plus cortical granules that have opened up immediately after fertilization.
(See Fig. 1.6.)

Structures similar to the fertilization membrane of echinoderm eggs
have been observed in other groups (*Ascaris*, frog, and trout). The
mechanism responsible for the raising of the fertilization membrane is
not clear. The explosion of cortical granules into eggs of echinoderms and

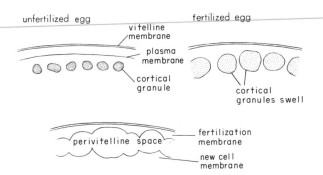

Fig. 1.6 Some of the changes taking place in the sea urchin egg immediately after fertilization. The vitelline membrane separates from the egg surface. The cortical granules below the plasma membrane swell and open up. The liquefaction of some of these granules produces vacuoles underneath the cell surface. This is the perivitelline space. The cell membrane of the fertilized egg is really a new plasma membrane formed from previous deeper-lying cytoplasm. It contains no portion of the previous oocyte plasma membrane, which has become incorporated into the fertilization membrane of the fertilized egg. The fertilization membrane is a fusion product consisting of the membrane and the vitelline layer. This newly formed structure constitutes the major barrier against multiple sperm invasion.

of many other phyla, an event that has been observed to occur within split seconds following sperm entrance, could be responsible for all these changes (See p. 25).

Rothschild and Swann[222] have shown that in sea urchin eggs one can distinguish a rapid change (two seconds) in which some, though not complete protection is provided against polyspermy and a slow change (established within two minutes), after which protection is complete. The slow change coincides with the explosion and disappearance of cortical granules, the raising of the fertilization membrane and a transient change in membrane potential and increase in K^+ ion exchange.

Monroy[198] has shown that sea urchin oocytes, unlike ova, can be penetrated by more than one sperm. Under polarized light, one of the differences between oocytes and mature ova that was noted was that the former lacked cortical *birefringence* but that in mature ova birefringence was present. Since birefringence is an indicator of orderly molecular rearrangement, it is, therefore, reasonable to argue that some subtle new arrangement of egg membrane molecules must take place and help guard the egg.

The role of cortical granules

If we look for a basic process underlying the elaboration of such diverse polyspermic blocks as the fertilization membrane, jelly coat, and the changed plasma membrane, it is probably to be found in the *cortical granule rearrangement* or *cortical granule explosion*, which seem to be the universal event immediately taking place in all eggs upon fertilization or parthenogenesis.

The presence near the egg's surface of cortical granules which upon fertilization are either expelled in order to participate in the formation of a membrane like the fertilization membrane of sea urchin eggs or the jelly coat of the frog egg, or simply explode to release a substance that may change the permeability of the egg plasma membrane, is so universal an occurrence among animals that one is justified in ascribing to it fundamental importance. That cortical granules and the material released by their explosion carry out an important function in the protection against polyspermy is demonstrated by a study comparing granules existing in the maturing oocyte with those present in the mature ovum of the sea urchin. In fixed and stained preparations of oocytes, the granules are shown to be scattered in the interior of the egg cytoplasm. They become concentrated in the egg cortex as the egg matures. There seems to be a definite correlation between the accumulation of granules in the egg cortex and the emergence of protection against polyspermy. In eggs of species in which the granules become established and concentrated at the cortex very early in the process of maturation, protection against polyspermy also emerges early. Conversely, in eggs of species in which the granules take their place relatively late during maturation the emergence of the polyspermy-preventing mechanism is delayed.

Experimentally, penetration of an egg by many sperm can be accomplished by subjecting the egg to a variety of treatments, all of which share the ability to inhibit the opening and explosion of cortical granules. Substances like 2,4-dinitrophenol, lactic acid and butyric acid tend to break down the polyspermy-prevention mechanism, presumably through their effect on cortical granules. All of these observations suggest strongly that the cortical granules upon explosion release some factor that speedily acts upon the plasma membrane to stimulate the establishment of a new and powerful block to further sperm entry.

The zona reaction

There is no phenomenon in mammals comparable to the raising of the fertilization membrane. Mammalian eggs are surrounded by a barrier of their own, the zona pellucida (see Fig. 1.1), whose structure appears to undergo a subtle change after sperm penetration. This change was termed the *zona reaction* by Braden, Austin, and David.[44] Occasionally, however, a second spermatozoon does succeed in passing through the zona pellucida of the rat egg. In such dispermic eggs, Braden, Austin and David observed that the relative position of the second sperm was always at the opposite pole to where the first sperm had entered. In rat eggs, the exact point of entrance of the sperm can be determined with relative ease because a minute slit appears in the zona pellucida, resulting probably from proteolytic digestion by acrosomal enzymes (See Chapter 2, p. 21). Based on these observations Braden, Austin and David suggested that polysperm invasion in mammals may be prevented by a structural change of the zona pellucida, the so-called zona reaction which is propagated from the point of entrance of the sperm and proceeds in both directions. This change takes longest to reach the opposite side, so that any second sperm that gains entrance will probably be admitted to the egg at a point directly across from where the first sperm had entered.

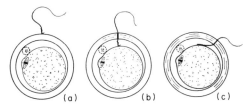

Fig. 1.7 Diagrams of rat eggs to show how the zona reaction is believed to spread out in relation to the point of sperm attachment on the surface of the vitellus. (From Austin[5])

By analogy with events observed in the eggs of sea urchins in which this problem has been more extensively investigated, one might postulate that the change in the zona pellucida might come about by secretion of substances from the egg interior upon fertilization. On this assumption, the attachment of the first fertilizing sperm to the surface of the plasma membrane of the eggs might cause, within split seconds, the release of a substance that diffuses through the perivitelline space and renders the zona pellucida impenetrable to any additional spermatozoa. This hypothesis is not entirely unsupported by observations. Disappearance of cortical granules from mammalian eggs has been described in the hamster by Austin[6] and the rabbit by Hadek.[127] Another bit of circumstantial evidence for the disappearance of cortical granules from fertilized mammalian eggs comes from observations that mammalian eggs of several species undergo considerable decrease in volume immediately after fertilization with a comparative widening of the perivitelline space.

The conclusions to be drawn from all this are summarized by Austin[6]:

'The best explanation of the mechanism of the zona reaction seems to be that the attachment of the fertilizing spermatozoa to the vitelline surface causes the release of a substance which diffuses through the perivitelline fluid and renders the zona pellucida impermeable to spermatozoa . . . '

This theory invokes a system that is widespread in the animal kingdom:

'. . . the arousal by sperm penetration of a reaction that is propagated over the egg surface and is associated with the release of an agent that has the function of rendering a membrane impermeable to spermatozoa.'

2

Penetration and Activation of the Egg

SPERM PENETRATION

The acrosome reaction

Morphology of acrosomal changes during fertilization

Penetration of the egg by the sperm is initiated by a so-called **acrosome reaction** which takes different forms in different species (Dan,[70–72] Dan et al.,[73] Colwin et al.[62]). The central part of the acrosome elongates into a tube which protrudes from the head of the spermatozoon. The significance of these acrosomal changes has been elucidated by the observations of the Colwins[62, 63, 65] on the penetration of the spermatozoon into eggs of the polychaete *Hydroides*, and the hemichordate *Saccoglossus*. Based on electron microscopic studies of fertilization in these two species, the Colwins have described a series of events that take place in the sperm (See Figs. 2.1. and 2.2.).

'On contact with the egg the tip of the sperm opens, and at the rim of the opening the acrosomal membrane fuses with the sperm plasma membrane. Thus the **acrosomal vesicle** is opened. As a result the contents of this vesicle, including granules of presumably lytic enzymes are released. The inner portion of the **acrosomal membrane** now everts and lengthens to form a tubular structure, the acrosomal tubule, which provides a tunnel through which the internal organelles of the sperm cell migrate [or better, must pass] to enter the egg.'

These events will now be discussed in more detail.

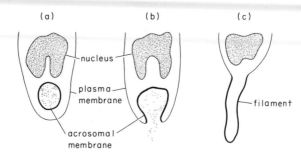

Fig. 2.1   The acrosome reaction in echinoderm sperm. (**a**) Sperm with intact acrosome surrounded by own acrosomal membrane. (**b**) Opening of acrosome and release of acrosomal granules. The acrosomal membrane and sperm plasma membrane have broken. (**c**) The acrosomal and plasma membrane heal and join. The new-joined membrane everts to send out the acrosomal filament. (Freely drawn after Colwin and Colwin[67])

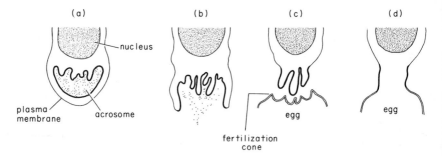

Fig. 2.2   The acrosome reaction in *Hydroides hexagonus*. (Freely drawn after Colwin[64])

Evidence for the presence of acrosomal lysins

The electron micrograph observations of Colwin and Colwin have been interpreted by them to mean that the granular material released from the acrosome might contain some lytic enzyme that can digest the egg barriers. Such a suggestion is certainly in accord with the developmental origin of the acrosome which is derived from the Golgi body of the spermatid cell. Golgi material in animal cells is usually associated with secretory activity. The contribution of **lysosomes**, generally regarded as containers of lytic enzymes in animal cells, to the formation of the acrosome deserves to be investigated also. The suggestion that one of the mechanisms employed by the sperm to penetrate the egg involves chemical digestion of the egg's outer barriers started a hunt for lysins in the acrosome (Tyler,[259] Berg[24]). Considerable evidence for the presence of lysins in the acrosome and for the role of the acrosomal lysins in sperm penetration has been accumulated for some species, but evidence for their presence in many other species is circumstantial and considerably debated (Dan[72]). For a review see Monroy[197] and Dan.[73]

A protein distinct from sperm antifertilizin has been extracted from limpet spermatozoa by Tyler.[259] The unfertilized egg of this species is surrounded by a tough coat which is not dissolved by concentrated acid applied over a period of hours, but suspension of homologous spermatozoa dissolves the coat in a few minutes. The lytic agent proved to be enzymatic by virtue of the fact it was nondialysable and thermolabile and gave a protein reaction. Spermatozoa of some sea urchin species have yielded membrane lysins active on the jelly coat or vitelline membrane or both. Berg[24] found similar lysins in seawater extracts of frozen thawed spermatozoa of *Mytilus*.

Tyler[261] concluded that such lysins are instrumental in enabling spermatozoa to penetrate through the membrane barriers to the surface of the egg cell proper. Proof that sperm make their own passageway by means of lysins has been provided by Colwin and Colwin.[67] The presence of lysins does not necessarily imply this function, however. Some of the lysins may be autolytic and may not normally attack the egg coats. On the other hand, failure to extract lysins from sperm of some species may not necessarily imply their absence. Their activity may depend on the presence of specific substances or on conditions that are satisfied only during egg-sperm contact.

Acrosomal function in mammalian sperm

The mammalian sperm becomes capable of fertilizing the mammalian egg only after staying for a period of time in the female genital tract. It is thought that during this period the spermatozoon undergoes a preparation referred to in the older literature as 'capacitation'. 'Capacitation' of mammalian sperm turned out to be a modification of the acrosomal reaction. These changes have been described and analysed by Austin,[5] Austin and Bishop,[8] Hadek,[127] and Srivastava, Adams and Hartree.[247]

Numerous attempts have been made to extract lysins from the spermatozoa of mammalian sperm.[127, 247] Austin[4] first provided evidence that hyaluronidase is present in the acrosomal portion of mammalian sperm and suggested that this enzyme may serve the function to disperse the cells of the corona radiata. However, a different enzyme is probably involved in preparing the sperm for penetration of the zona pellucida, for the zona is composed of mucoproteins, and is, therefore, not altered by hyaluronidase. Breakdown of the zona must consequently be accomplished by proteases (Srivastava *et al*[247]). Narrow slits 1–2 μm in diameter have been observed by Austin[6] in the zona pellucida of rodent eggs

following sperm penetration, suggesting that the lysins might exert a
strictly localized action on the zona pellucida.

Gamete fusion

Mechanical models

The zoological literature antedating electron microscopy entertains
the vague notion that the sperm cell invades the egg and that this event is
followed by fusion of two nuclei. The cavalier attitude implicit in so
many textbooks describing egg-sperm association, as if the invasion of
one cell by another is the most commonplace event in the biological
world, is certainly amusing. The genius of Loeb, who over fifty years ago

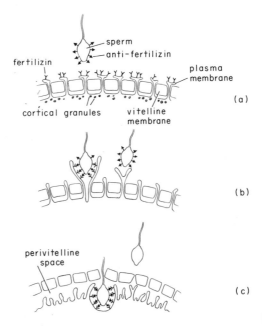

Fig. 2.3 Tyler's model of sperm entrance into the ovum. **(a)** The plasma membrane protrudes through pores of vitelline membrane. It bears specific receptor
sites (fertilizin) complementary to the antifertilizin of the sperm. **(b)** The plasma
membrane extends around the fertilizing sperm by progressive interaction of
fertilizin with the antifertilizin of the sperm. At the same time the plasma membrane
is retracted at other sites, shaking off all spermatozoa that are less tightly locked to
the fertilizin of the ovum, thus preventing polyspermy. **(c)** The plasma membrane
with the engulfed sperm withdraws through the pores in the vitelline membrane.
At the same time cortical granules explode, pushing the vitelline membrane up.
(After Tyler[264])

was the first to examine the concept of phagocytosis as a possible mechanism of sperm penetration, is all the more apparent.

According to more modern biological concepts, phagocytosis of large particles by cells is presumably accomplished by a membrane engulfing mechanism referred to as *pinocytosis*. Tyler[264] proposed a theoretical model whereby the peculiarities of the egg membrane could operate to draw in the sperm by such a pinocytotic process.

The main objection to Tyler's scheme seems to be that the spermatozoa with its cell membrane intact would remain separated from the rest of the egg cytoplasm. Some process would have to be invented that would effectively break down the plasma membrane of the sperm in order to allow it to interact with the egg cytoplasm.

Membrane fusion

The fusion process between the egg and sperm as revealed by the electron microscope studies already mentioned has been described in great detail by the Colwins.[62, 64–67]

1. Upon contact of the sperm apex with the outer part of the egg envelope, the apex of the sperm undergoes dehiscence, thus opening the acrosomal vesicle to the exterior. The acrosomal contents including the acrosomal granule, which probably contain egg envelope lysins, are thus brought into contact with the egg envelope. (Figs. 2.4(b) and 2.5(b).)

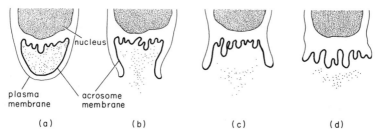

plasma membrane · nucleus · acrosome membrane

(a)   (b)   (c)   (d)

Fig. 2.4   Pattern of sperm-egg association in *Hydroides*. For explanation see the text. (Freely drawn after Colwin[64])

2. During the process of dehiscence, the apical portion of the sperm plasma membrane and the outer portion of the acrosomal membrane fuse around the rim of the opening resulting from dehiscence. (Figs. 2.4(b) and 2.5(b).)

3. The inner (adnuclear) portion of the acrosomal membrane begins to lengthen, forming the **acrosomal tubule**. As the acrosomal tubule lengthens, the remaining membrane of the acrosomal vesicle everts (so

the *now* apical part of the sperm plasma membrane consists of former acrosomal membrane).

4. The tubule extends through the egg envelope and makes contact with the egg plasma membrane. (Figs. 2.4(c) and 2.5(c).)

5. The tubule interdigitates slightly with the egg plasma membrane and then fuses with it, so that there is now one continuous (zygote) plasma membrane. (Figs. 2.4(d) and 2.5(d).)

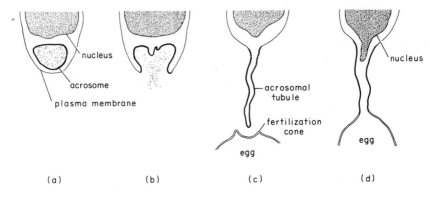

(a)   (b)   (c)   (d)

Fig. 2.5   Pattern of sperm-egg association in *Saccoglossus*. For explanation see text. (Freely drawn after Colwin[64])

6. As the result of the fusion of sperm and egg plasma membranes, a joint tunnel between the two cells is accomplished through which the nucleus and the cytoplasm containing the organelles of the sperm cell move across into the egg cytoplasm. (Figs. 2.4(d) and 2.5(d).)

The pattern of acrosomal preliminaries followed by **gamete-membrane fusion** first revealed by the Colwins for *Hydroides* and *Saccoglossus*, has also been described for members of other phyla, i.e., echinoderms, molluscs and vertebrates.

Possibly, membrane fusion constitutes one of those basic and fundamental biological mechanisms shared by the reproductive cells of all organisms.

In the Colwins' scheme, the role of the acrosomal region would, therefore, appear to be a multiple one, for the acrosome delivers the lysins that attack the outer egg envelope, while the tip of the acrosomal tubule, by fusing with the plasma membrane of the egg, establishes the zygote. Lastly, the suggestion has been advanced by the Colwins that the acrosome may deliver substances that are instrumental in activating the egg. The problem of egg activation will be discussed in the next section.

ACTIVATION OF THE EGG

Morphological changes following sperm contact

The simple observation that the act of fertilization changes the egg from a state of slow decline to a new spurt of activity has been interpreted by embryologists to mean that the sperm, in addition to delivering an extra set of chromosomes, activates the egg. (For a review, see Monroy and Tyler.[200])

On a morphological level, *egg activation* can be described as a series of surface changes that take place immediately following contact of egg and sperm. Some of these changes have been extensively described, such as the explosion of cortical granules, the raising of the fertilization cone, and the raising of the fertilization membrane (see Chapter 1, p. 14).

There is good evidence that the mere establishment of contact between the spermatozoon and the egg plasma membrane is sufficient to spark the activation reaction.

Lillie[62] showed that, in the egg of *Nereis*, attachment of sperm to the egg membrane is immediately followed by surface changes within the egg and the formation of the fertilization cone, while the actual penetration of the spermatozoon does not take place until 40 minutes later. Most probably, the spermatozoon establishes immediate contact with the plasma membrane through the acrosomal tubule and the activation of the fertilization reaction is thus elicited. Hiramoto[142] removed spermatozoa from sea urchin eggs at the start of fertilization and found that when the sperm is removed from the egg surface before it has entered the egg cytoplasm, some type of parthenogenic activation still follows (i.e. the egg undergoes a number of monoaster cycles without cleavage).

Metabolic changes

Changes in oxygen consumption following fertilization

A milestone in the discovery of the nature of the *egg activation reaction* was reached when Otto Warburg[277] demonstrated that within a few minutes of fertilization, the rate of O_2 consumption of sea urchin eggs increases several fold. This observation suggested what had been implied so long, that the mature unfertilized egg is a metabolically inhibited cell and that fertilization constitutes an act of *deinhibition*. Since this first discovery, a lot of labour has been invested in the study of the oxygen consumption of immature and mature eggs, chiefly by Borei,[27, 28] Brachet,[29] Horwitz,[156] Rothschild,[220] and Whitaker.[291, 292]

The conclusion reached from these studies was that during maturation of sea urchin eggs considerable metabolic inhibition builds up and that

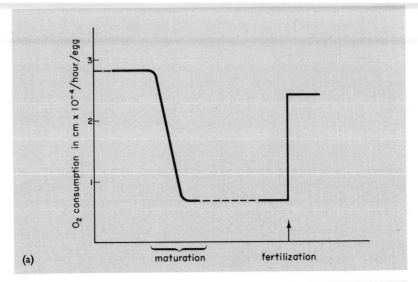

(a)

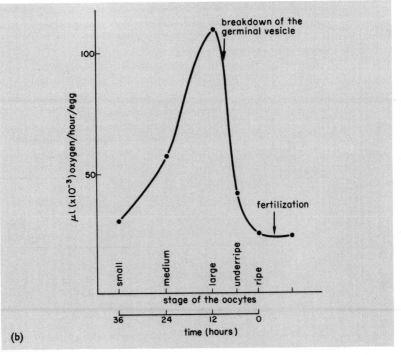

(b)

fertilization in some manner removes these hypothetical inhibitors, thus allowing for the new spurt of metabolic activity. Intuitively, one feels that profound changes in metabolism should occur at fertilization and that this may well be reflected in O_2 consumption or respiratory quotient (RQ) of the egg. Unfortunately, intuition cannot always be confirmed by experimental observations. The spurt in respiratory change observed in freshly fertilized sea urchin eggs, which so excited Warburg,[277] is not a universal occurrence. Studies on eggs of a number of other invertebrate and vertebrate species have shown that fertilization is indeed accompanied by an increase in respiration of some eggs but in others a decrease was noted, and still others showed no change at all.

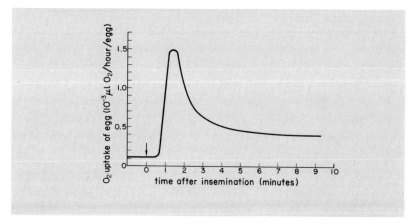

Fig. 2.7 Change in the rate of oxygen consumption of eggs of *Pseudocentrotus depressus* during the first 10 minutes following fertilization, measured polarographically. Arrow indicates the addition of sperm. (From Ohnishi and Sugiyama[215])

The conflicting observations may be reconciled by the assumption that the stage of maturation attained by the ovum at the time of fertilization may determine the direction and the magnitude of changes in respiratory metabolism. Whitaker[291, 292] has advanced convincing evidence in favour of the contention that the level of metabolism that had existed in the primary oocyte prior to the decline it suffered during subsequent maturation is re-established to the original level after fertilization. He argued

Fig. 2.6 (a) Changes in rate of oxygen consumption during maturation and at the time of fertilization in the egg of *Paracentrotus lividus*. **(b)** Rate of oxygen consumption in the course of maturation and following fertilization in the egg of *Oryzias latipes*. (Calculated from data of Nakano.) (From Monroy[198])

that the increase in respiration that follows fertilization may occur in only those eggs (like those of sea urchins and *Fundulus*) that have completed maturation at the time of fertilization. Eggs of other species that are penetrated by sperm at a time when nuclear and nuclear-dependent maturation has not yet been completed may not show any immediate respiratory increase following sperm entry. In fact, those eggs that still continue the maturation process while the sperm waits in the egg cytoplasm may continue the respiratory decline until their own maturation has been completed and they are ready for syngamy.

Mere descriptions of changes in oxygen consumption do not tell us much. It is important to know what they signify. The measurement of oxygen consumption is at best a diagnostic tool in the hands of the developmental biologist. It signals to the investigator that important processes are taking place or are in preparation, which deserve his attention, just as a change of temperature registered on the fever thermometer alerts the physician to seek the identity of the suspected viral or microbial troublemaker.

Changes in respiratory enzymes and accumulation of metabolic intermediates in fertilized eggs

It has been known for some time that in unfertilized sea urchin eggs there is a build-up of polysaccharides and a rapid breakdown of the polysaccharide fraction within minutes after fertilization. Correlated with this change is the appearance of considerable amounts of acid following fertilization. This increase in acid, referred to as **fertilization acid** in the older literature, and since identified as lactic acid, is precisely what one would expect if metabolic activity increases.

Other qualitative changes that have been reported include an increase in the level of hexose phosphates, which is very low in unfertilized eggs and rises considerably after fertilization. (For review see Monroy.[198])

Increased activity of G-6-P-dehydrogenase following fertilization of sea urchin eggs has been reported by Isono (Monroy[198]) and of 6-phosphogluconate dehydrogenase by Backstrom (Monroy[198]) and of aldolase by Ishihara (Monroy[198]). In the unfertilized egg, G-6-P-dehydrogenase is tied to a fraction sedimenting at low speed. Shortly after fertilization, the enzyme is found in the soluble fraction recoverable in the supernatant.

These findings may imply that the deinhibition of enzymes may be brought about by releasing the enzymes from specific cellular organelles or compartments in which they are 'incarcerated' and prevented from interaction with available substrate.

The possibility that metabolic inhibitors accumulate in the oocyte in the course of maturation and that they are got rid of at fertilization has been repeatedly suggested, particularly by Runnstrom[226] and later by Brachet.[29, 37] It was reinforced by the discovery by Maggio and Monroy (Monroy[198]) that extracts of unfertilized sea urchin eggs exert a strong inhibition on the activity of cytochrome oxidase of liver mitochondria, whereas extracts of fertilized sea urchin eggs lack this inhibitory action. Brachet[37] has suggested that metabolic inhibitors may be got rid of more easily after fertilization because of sudden changes in permeability. If that is the case, the inhibitor must be either extramitochondrial or diffusable from mitochondria. This assumption is strengthened by the observation that mitochondria obtained from unfertilized or fertilized eggs exhibit *in vitro* the same cytochrome oxidase activity.

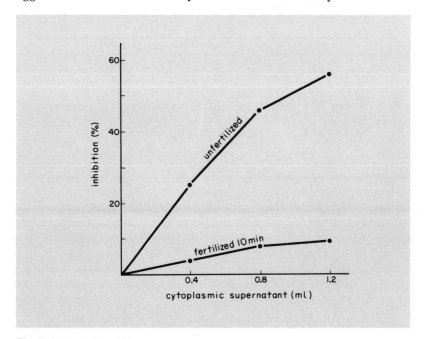

Fig. 2.8　Inhibition of cytochrome oxidase activity of rat liver mitochondria by different concentrations of 105,000 g cytoplasmic supernatant of unfertilized and fertilized eggs of *Paracentrotus lividus*. Both supernatants had the same concentration. (From Maggio and Monroy, in Monroy[198])

Effect of fertilization on protein synthesis

Does fertilization affect incorporation of amino acids?

The next question that logically suggests itself is whether the changes in respiratory metabolism reflect changes in protein synthesis. Assuming

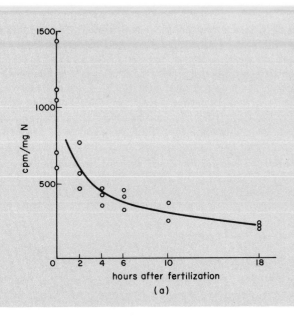

hours after fertilization

(a)

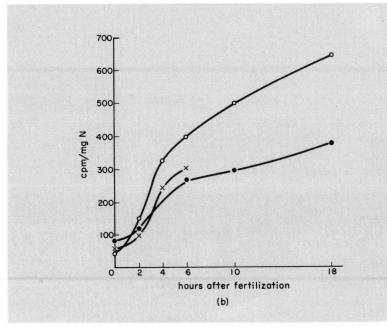

hours after fertilization

(b)

that amino acid incorporation is a good indicator of protein synthesis, it was therefore of interest to find out whether rate of amino acid incorporation of eggs is enhanced at fertilization. Experiments with homogenates from fertilized eggs incorporate more than twice as much labelled amino acids as do unfertilized eggs (Hultin in Monroy,[198] Nakano et al.[203]). Since all determinations were carried out on homogenates, permeability could not have been a deciding factor. Additional evidence that the sudden change in the rate of incorporation observed at fertilization does not depend on greater access of the egg to the supply of amino acids comes from experiments by Brachet (Monroy and Tyler[200]) who loaded unfertilized eggs with radioactive amino acids by microinjection and noted no incorporation into proteins. Nakano and Monroy[204] came to a similar conclusion based on experiments in which labelled amino acids that were administered to the mature female failed to be incorporated into proteins of the ova.

Are ribosomes activated in the fertilized egg?

The question arises which component of the protein-making machinery of the cell is not functioning properly prior to fertilization. The first clue as to where the gap might be located came from experiments by Hultin and Bergstrand (Monroy[198]) who showed that in a cell-free system ribosomes prepared from unfertilized eggs exhibit little ability for incorporation of amino acid into proteins, while ribosomes obtained from freshly fertilized eggs are quite active in carrying out such incorporation. This could mean that the gap in the machinery of protein synthesis of unfertilized eggs is located at the ribosomal level. Hultin[159] interpreted these results to imply that upon fertilization ribosomes undergo a structural rearrangement whereby they become activated. This, however, could not be the whole story for the following reason. Students familiar with molecular biology may recall that Nirenberg, the American Nobel Prize winner for medicine, demonstrated that polypeptides could be linked up in a cell-free system placed in a suitable medium containing all the amino acids and messenger RNA.

The addition of a synthetically polymerized messenger RNA consisting only of the nucleotide uracil to such cell-free systems stimulated the synthesis of a polypeptide chain composed exclusively of phenylalanine.

Tyler,[265] Nemer,[208] Nemer and Bard[209] and Wilt and Hultin[296] showed that ribosomes that were removed from unfertilized sea urchin

---

Fig. 2.9 Unfertilized eggs of *Paracentrotus lividus* were prelabelled with ^{35}S-methionine by injection into the body cavity of the female; on the ordinate, radioactivity in the unfertilized eggs. (a) The radioactivity of the TCA-soluble fraction decreases rapidly after fertilization, while at the same time (b) it increases in the mitochondria. (From Nakano and Monroy[204])

eggs, and placed into a cell-free system containing synthetically made polyuridylic acid (poly u) mRNA, stimulated incorporation of phenylalanine into a polypeptide chain. This finding means that the missing link in the unfertilized egg cannot be the availability of functional ribosomes.

The assumption that the cytoplasm of the unfertilized egg contains special inhibitors of protein synthesis must be rejected too. It has been shown that the cytoplasm of unfertilized eggs can serve as a medium in which incorporation of labelled amino acids by ribosomes can take place *in vitro* provided that mRNA is present. Addition of cell sap from unfertilized eggs did not interfere with active incorporation of labelled amino acids.[159]

Do unfertilized eggs lack messenger RNA?

An alternative suggestion might be that the unfertilized egg lacks messenger molecules. This sounds like a reasonable hypothesis. However, other considerations make it highly unlikely because considerable evidence has been accumulated that mRNA which programmes the fertilized egg during the initial phase of development must have been prepared prior to fertilization and is kept in storage.

The demonstration of existing stable templates in unfertilized eggs was first presented by Gross and Cousineau.[116, 117] These authors have shown that exposure of fertilized eggs to **actinomycin D** while shutting down the synthesis of new mRNA did not interfere with the initial development of the egg. Cleavage and blastula formation proceeded quite normally in the presence of this inhibitor of RNA synthesis and no ill effects were observed until gastrulation. Since development up to the gastrula stage must require the synthesis of some new proteins, they concluded from these observations, that the fertilized egg carries out its protein synthesis during cleavage with the help of mRNA molecules that have been synthesized prior to fertilization and that the mRNA which programmes the fertilized egg during the initial phase of its development was probably prepared in the oocyte.

A more sophisticated demonstration for the existence of stable templates in unfertilized eggs, is based on the following strategy. Failure to find RNA labelled with radioactive precursor after fertilization is necessary but not sufficient evidence that RNA is not being synthesized. The pools of the immediate precursor might not be available to exogenous label. More useful would be a system in which mRNA synthesis could be demonstrated. If this synthesis could then be turned off without causing accompanying depression of protein synthesis, it could be concluded that the proteins were assembled independently of the simultaneously occurring genetic readout of messenger molecules formed. If, under those conditions protein synthesis continues while the mRNA synthesis stops, one could conclude that pre-existing messengers provided the templates for protein synthesis. Such a system has been found and studied by Gross, Malkin and Moyer.[118] Eggs of two species of sea urchins, *Arbacia punctata* and *Lytechinus*, were fertilized; aliquots were then removed at intervals to test tubes containing 0·05 ml of ^{14}C-labelled valine. After 20 minutes, incorporation was stopped by adding trichloracetic acid (TCA). After centrifuging, proteins were precipitated and prepared for scintillation counting.

In controls, protein synthesis began at fertilization. The rate of synthesis grew steadily for about 3 hours; it remained constant for 8 hours post fertilization. Incorporation increased rapidly again for the next 7 hours until it reached a plateau at 15 hours post fertilization. In an experimental series of eggs which had been exposed to 20 μg of actinomycin D for 3 hours prior to fertilization, the post-fertilization release of protein synthesis took place exactly as in controls. By 5 hours, the rate was somewhat higher but embryos treated with actinomycin D did not show the second increase that was so noticeable in the controls and subsequently protein synthesis began to decline. It would appear that the suppression of synthesis of mRNA that supposedly follows actinomycin D treatment operated to reduce the second rise of protein synthesis in the egg but not the first. The initial rise remained unaffected by the interference with messenger synthesis taking place prior to fertilization. This suggests that early protein synthesis must have relied on messengers that were already produced prior to the interference by actinomycin D. (See Fig. 5.4.)

Failure of the second rise of protein synthesis to occur in eggs treated with actinomycin D probably reflects absence of an important new RNA that has not been elaborated in embryos similarly treated. (For a review see Monroy and Tyler,[200] Gross[114] and Tyler.[267]) The relevance of this finding to the present discussion consists in the demonstration that mRNA must be present in the unfertilized egg and that the low rate of protein synthesis in unfertilized eggs is not a reflection of absence of messenger RNA in their system prior to fertilization.

Unmasking the messenger?

If it is true that the inability of the unfertilized egg to promote protein synthesis is not a reflection of the lack of functionally capable ribosomes nor of the availability of messenger RNA molecules, the third alternative that suggests itself is to assume that the mRNA that is stored in unfertilized eggs is prevented from combining with the ribosomes to form a collaborating functional system. Under this assumption the unfertilized egg contains both ribosomes as well as messenger RNA which need to be brought together in order to carry out protein synthesis. Possibly, the two components of the machinery are kept apart in the unfertilized egg and are brought together as the result of fertilization. The early changes taking place in most eggs which even the naked eye can recognize as a turbulent movement may well accomplish the rearrangement of particles which may bring ribosomes and messengers into spatial proximity necessary for them to function. A more sophisticated proposal was advanced by Spirin.[246] He suggested that perhaps mRNA that is stored in the unfertilized egg is present in 'masked' or unusable form and that the reason for the low level of protein synthesis of unfertilized eggs is the masked state in which its mRNA is stored. Consequently, the ribosomes of the unfertilized eggs are not programmed until the fertilizing sperm removes the inhibition and releases the mRNA from its inhibitor. Evidence in support of this theory is reviewed by Spirin[246] and is based on results of a search for particles of RNA other than ribosome or polysomes in egg cytoplasm. Particles of protein-coated RNA have been prepared as pellets from microsomes of sea urchin eggs. There is some indication that at least two forms of bound mRNA exist in fractions of unfertilized eggs of the teleostan fish, *Misgurnus fossilis*, and that they may be present in a variety of other cell types. The stable form of bound mRNA

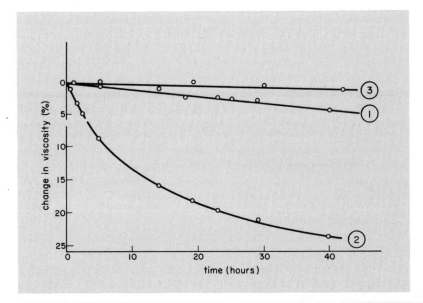

Fig. 2.10 Proteolytic activity of extracts of *Paracentrotus* eggs 3 minutes after fertilization (2) as compared to that of unfertilized eggs (1) and of eggs 30 minutes after fertilization (3) as measured from the viscosity-lowering effect on gelatin. (From Lundblad[188])

forms a series of discrete components with sedimentation coefficients of 20–70 S constituting an mRNA protein complex, that has been referred to in the literature as the so-called *informosome*. In order to be available for protein synthesis, such a particle must have its protein coat removed. The existence of a second form of inactive mRNA that is already attached to ribosomes has been demonstrated in unfertilized sea urchin eggs by experiments of Monroy, Maggio and Rinaldi (Monroy and Tyler[200]). Treatment of these ribosomal pellets with trypsin transformed inactive preparations into active ones that were fully capable of promoting protein synthesis without addition of extraneous mRNA. The trypsin treatment presumably removed the protein pellet from these informosomes, thus providing a working RNA polysome complex. (For a review, see Spirin.[246]) Hence, the possibility exists that during oogenesis, informosomes are formed and become attached to ribosomes, with their protein cover intact. The removal of this protein by the fertilizing stimulus might liberate the mRNA.

The old discovery by Lundblad[188, 189] that proteolytic activity increases within minutes after fertilization of sea urchin eggs might fit in well with this theory, for these enzymes might be instrumental in the unmasking process that transforms informosomes into active mRNA polysome complexes. Once this is accomplished, protein synthesis is switched on, and the drama of embryogenesis can begin to unroll.

3

Theories of Induction

THE DISCOVERY OF THE ORGANIZER

The pioneers: Weismann, Roux and Driesch

If, as pointed out in the preceding chapter, the sperm acts entirely unspecifically, then the egg must contain all the information and all the equipment necessary for developing an embryo and subsequently a miniature adult. The only reasonable hypothesis is to assume that the egg must be preformed and preorganized. In the 1880s, the German zoologist August Weismann had developed the so-called germ-plasm theory, which pictured the egg as a *mosaic* containing specific organ-forming materials. All that happens in subsequent development, according to Weismann, is a kind of sorting out process through which the highly organized egg, by a series of cell divisions, distributes the different organ-forming materials and parcels them out to different cells. In 1888, Roux,[224] an anatomist at the University of Breslau, performed a classical experiment to test this theory.

He took a two-cell stage of a frog egg, killed one cell with a hot needle, and found that the other cell developed into only half a tadpole. Clearly, the frog egg was a mosaic, each cell carrying one half of the machinery for constructing a frog. During cell division, the genetic material is simply distributed. The mosaic theory had been justified and it seemed that experimental embryology had solved one of its problems before it had gotten started in earnest. However, there remained a lingering doubt. Since the killed cell had remained attached to the living one, it might have influenced the living partner. Driesch,[81, 82] therefore, tried a somewhat different approach. He separated single cells of 2, 4, 8 and 16 cell stage sea urchin blastulae of *Echinus microtuberculatus* and placed them into vials of sea water without injuring them. Each single blasto-mere cell developed into a whole larva.

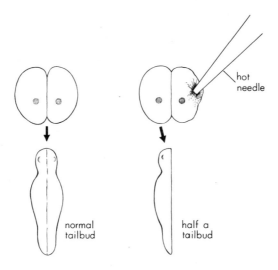

hot needle

normal tailbud

half a tailbud

Fig. 3.1   Roux's experiment. One blastomere of a two-cell stage of a frog egg was killed with a hot needle. The surviving cell gave rise to a half embryo. The mosaic theory had scored a point.

Was this just a peculiarity of an *Echinus* cell? Driesch[81] was the first to speculate that it might be a far more universal phenomenon. 'When compared with Roux's, my results reveal a difference in behaviour in the sea urchin and frog, yet perhaps this difference is not so fundamental after all. If the frog blastomeres were really isolated and the other half (which was probably not dealt with in Roux's case) had really been removed, would they not perhaps behave like my *Echinus* cells?' Apparently, embryonic cells act as part of a population when part of a blastula, but are able to form complete embryos, when separated and single. Something in the changed environment must signal to these cells that they are no longer part of the whole, but on their own, and must induce them to carry out all the processes necessary to form whole embryos. The idea that embryonic cells 'understand' and that they can 'interpret' and act upon signals of their immediate environment was new, dramatic and sensational and provided the starting point for the fundamental experiments of Hans Spemann. Working with eggs of the

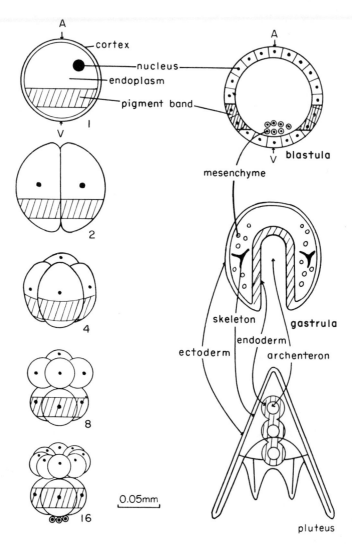

Fig. 3.2   How the contents of the egg may be traced into the embryo (pluteus). Upper left diagrams the egg: A = animal pole; V = vegetal pole. Following down left side are the 2-cell, 4-cell, 8-cell and 16-cell stages. Top right shows a section of a blastula with the cells from the vegetal pole entering the cavity of the blastula. Immediately below is a gastrula with the archenteron. Finally, we see the pluteus with the pigment band localized in the gut. (From Barth[15])

newt *Triturus,* he separated the two cells of the first cleavage stage by simply constricting the egg with a strand of baby hair across the middle before it cleaved, leaving only a thin bridge of protoplasm between the two halves.

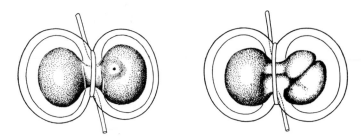

Fig. 3.3   Classic experiment performed by Hans Spemann. A salamander egg was tied into two halves across the grey crescent. The nucleus of the egg (nuclei at right) was confined to one half. Initially only the half containing the nucleus divided. (From Gray[100])

The side containing the nucleus began to divide into 2, 4, 8, and 16 cells. The other side remained single until one of the neighbouring cells was close enough to the pinched waist to send its nuclear material across the bridge. Eventually, the other half caught up and two complete animals developed. When next he pinched the cleaving egg in a direction at right angles to the constriction tried before, a result very different from the preceding one was obtained. The delayed, uncleaved half, instead of forming a normal looking newt, gave rise to an unorganized **Bauchstück**

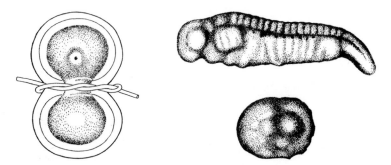

Fig. 3.4   Second experiment by Spemann. The egg was tied parallel to the grey crescent. The half containing the grey crescent developed into a normal embryo, whereas the other half produced only an unorganized 'belly piece' or 'Bauchstück'. (From Gray[100])

or **belly piece**. It contained liver, gut, and other endodermal organs but no neural or sensory structures.

One of the peculiarities of the newt egg is that immediately after fertilization, a small crescent-shaped segment on the surface of the egg takes on a greyish appearance. This is the **grey crescent.** The only apparent difference between Spemann's two experiments was that in the second attempt, not only the nucleus, but also the grey crescent was excluded. Since only the cell containing the grey crescent developed into an

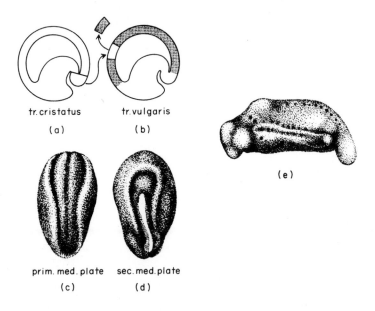

tr. cristatus   tr. vulgaris

(a)   (b)

prim. med. plate   sec. med. plate

(c)   (d)

(e)

Fig. 3.5   (a) Heteroplastic transplantation of the blastopore lip of *Triturus cristatus* gastrula (colourless) to the (b) gastrula of *Triturus vulgaris* (pigmented). Note that under the influence of the additional blastopore lip a secondary embryo is formed composed of the cells (pigmented) of the *vulgaris* host. (From Saxen and Toivonen[236])

embryo while the half lacking it formed an unorganized mass, Spemann began to wonder what was so significant about the 'grey crescent'. In fact, this crescent is only a surface feature but it defines the position of the lip of the **blastopore** through which the cells roll in during gastrulation, and it was to this lip that Spemann directed the attention of his pupil, Hilde Mangold.

Spemann's and Mangold's classic experiments

Spemann suggested that a heteroplastic transplant should be made and the dorsal lip from the gastrula of *Triturus cristatus*, a colourless species of newt, be grafted into the blastocoel of *Triturus taeniatus*, a dark pigmented species.[243]

Under the influence of two blastopore lips, the embryo developed into a double-headed monster with two axial systems, including two heads, two brains, and two whole neural cords. The secondary embryo had all of its organs built from cells derived from the host, i.e. dark cells. This clearly indicated that the transplanted dorsal lip did not contribute any cell itself, but merely organized and directed the formation of the secondary embryo. Spemann called the dorsal lip an **organizing centre** or **organizer** and suggested that the cells of the blastopore that subsequently roll in to form the roof of the archenteron induce the overlying ectoderm into neural structures. To test this proposition, he removed the dorsal lip of salamander gastrulae and noted that embryos deprived of their dorsal lip failed to form either a neural tube or head structures. These results convinced him that the dorsal lip and the roof of the archenteron induce the overlying sheet of ectoderm to roll up into a tube and differentiate into neuroblasts.

Prospective potency and prospective fate

The results described above indicated furthermore, that embryonic tissues can and do respond to cues or signals. Whether such signals are really necessary prerequisites for embryonic development under conditions prevailing in nature, is another question. The proposition that the egg is responsive to cues from surrounding cells was tested by a variety of transplantation experiments, to be described below. Vogt,[269, 270] followed the distribution of vital dye particles that had been placed into cells of amphibian blastulae until the tailbud stage was reached. On the basis of these observations he constructed a *fate map* of the blastulae, in which the geographical boundaries of organ-forming sites, such as somite, spinal cord, eye, brain, gut and brain-producing area could be clearly delineated. With this method the *prospective fate* of each cell of the blastula could be identified. Interference with this system, such as removal of cells of a given organ-forming area and transplantation into a different site, will turn development into unorthodox directions.

For example, cells that have been removed from the prospective eye region of the blastula and implanted into the tail or developing limb of a tailbud stage can apparently be switched into a very different path by influences exerted on them by the environment. It is as if, in their new

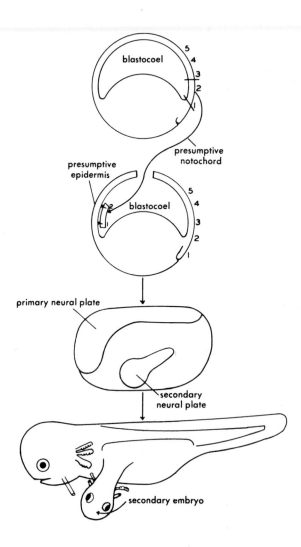

Fig. 3.6 The formation of a secondary head after transplantation of regions 1 and 2 in the early gastrula. Regions 1 and 2 are transplanted (top) into another gastrula in such a way as to bring them into contact with the presumptive epidermis. The second gastrula forms a primary neural plate and a secondary neural plate. The final result (bottom) is an embryo with two heads. It is important to note that the secondary embryo contains only a very small amount of tissue contributed by the transplant. Most of the tissue comes from the host. (From Barth[15])

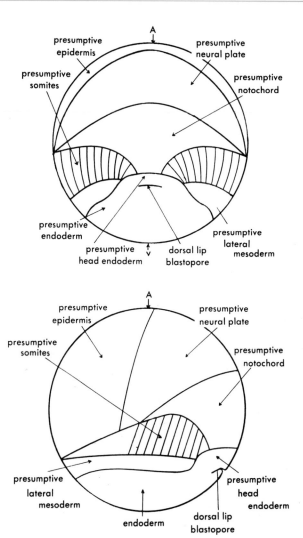

Fig. 3.7   Maps of the presumptive values of the regions on the surface of the early amphibian gastrula. The maps of various species of amphibians are different, and this diagram merely gives relative positions and sizes of the presumptive structures. Top: a top view of the early gastrula with the dorsal lip as a point of orientation. Bottom: a side view of the same. (From Barth[15])

surroundings, the prospective 'eye' cells are being instructed to perform a different task from the one they normally perform and are thus pushed to develop into muscle, cartilage and bone of tail or limb rather than into retina or lens. In the blastula, the ***prospective potency*** of cells seems to exceed the prospective fate. When the same experiments were tried with gastrula cells instead of with blastomeres, the results were radically different. Cells that had rolled over the lip of the blastopore no longer responded to signals when transferred to new surroundings. Prospective eye cells removed from the overlying ectoderm of a gastrula and transplanted to the dorsal axis of a tailbud no longer formed caudal structures. Instead, an embryo equipped with eyes on its rear end emerged. The

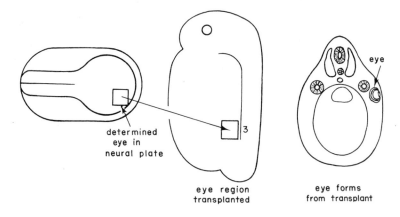

determined
eye in
neural plate

3

eye region
transplanted

eye forms
from transplant

eye

Fig. 3.8 The presumptive eye at the neurula stage forms only an eye when transplanted. The presumptive eye is cut out of the neural plate of a neurula and is transplanted to position 3 in an older embryo. It develops into an eye, whereas in Fig. 3.9 the presumptive eye of the early gastrula forms notochord, somites, pronephros, and spinal cord. (From Barth[15])

results of these experiments have been interpreted to mean that after gastrulation prospective potency no longer exceeds prospective fate, but becomes very restricted.[241] The most important lessons learned from them, however, is the conclusion that signals are exchanged between parts and cells of embryos, and that such signals are apparently required in order to allow differentiation to proceed. Removing the transmitter of the signals as was done in the classical experiment by Spemann,[242, 243] who extirpated the dorsal lip and the archenteron, effectively stops further differentiation of the ectoderm.

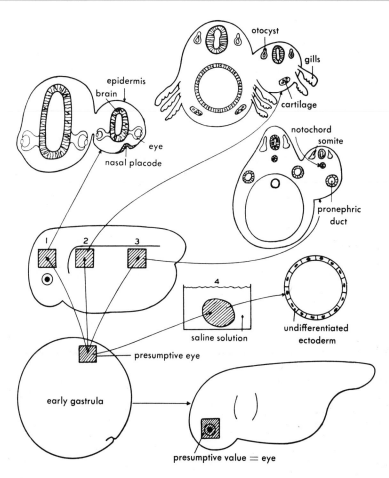

Fig. 3.9 A test for the potencies of the presumptive eye of the early gastrula with some of the results. A region of the early gastrula, at lower left, is stained and found to develop into the eye, lower right. The presumptive eye is cut out and transplanted to region 1, region 2 and region 3 of an older embryo. In region 1 the presumptive eye develops into epidermis, brain, nasal placode and eye (upper left). Finally, in region 3 somites, notochord, pronephric duct and spinal cord develop out of tissue which normally forms eye (right centre). If the presumptive eye is placed in saline solution it forms a ball of undifferentiated ectoderm. (From Barth[15])

THE QUEST FOR THE IDENTITY OF THE INDUCTOR

Extracting the inductor

A priori one might assume that nothing would be easier than silencing the emitter of the signals. Bautzman, Holfreter, Spemann and Mangold[18] demonstrated that contrary to expectations, killing the dorsal lip and archenteron does not silence the inductor. This finding led to the hypothesis that the message transmitted by the inductor must be spelled out in chemical terms and therefore can probably be just as easily released from freshly killed as from living tissue. An intense search for the chemical identity of the inductor started. Bautzmann subsequently showed that the organizer may be killed by a variety of means like crushing, freezing, or heating, and by treatment with different kinds of solvents like alcohol, ether and chloroform. After being subjected to all these treatments the organizer tissue can still induce. It might be said that Bautzman's experiments cut short the 'Age of Innocence' of the 'Golden Period' of Embryology that was inaugurated by Spemann and his colleagues. Spemann[242] himself had discovered that the power of neural induction was shared by tissues other than the archenteron and dorsal lip. Holtfreter[143] had demonstrated that a large variety of vertebrate and invertebrate tissues can induce neural tubes and that induction was not a property residing only in living tissue but could be found in killed, boiled, minced, and fixed tissues as well. In fact, extractants like formaldehyde can induce as effectively as does living tissue. Furthermore it was discovered that almost anything and everything, from dead tissue to turpentine-soaked cotton plugs, had the power to induce the overlying ectoderm to roll up into a tube. The only possible generalization emerging from these experiments was that induction is possibly accomplished by a substance which must be very widely distributed among practically all of the known adult and embryonic tissues, and that the archenteron holds no monopoly for this agent. In early embryos lacking a circulatory system, the archenteron roof probably serves as the source of this material since it fulfills the requirement of proximity allowing for easy diffusion to the target, i.e. the overlying ectoderm. And so the search for the inductor lasted all through the 1920s and 1930s.

First attempts at identifying the inductor

Needham, Waddington and Needham,[207] on the basis of certain extractions made from the amphibian neurula, suggested that the inductor might be a steroid. A German group, led by Lehman,[182] came to the conclusion that the inductive stimulus was the property of certain organic acids. Other inductively active compounds such as cephalin were found by Barth,[16] and methylene blue by Waddington, Needham and Brachet (quoted in Saxen and Toivonen[236]). There were just too many tissues, extractants, substances, and treatments available that could induce ectoderm to neuralize, to find any common denominator

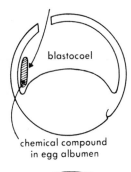

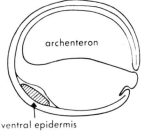

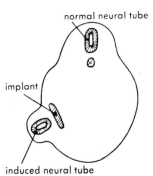

Fig. 3.10 Chemical compounds induce neural tubes. The chemicals are dispersed in egg albumen and the mixture is placed in the blastocoel of an early gastrula. At the end of gastrulation the implanted mixture is found in contact with the ventral epidermis. Later a cross-section of an embryo reveals an induced neural tube. (From Barth[15])

between them. The only theory that made sense to explain the great variety of active inductive agents was the one advanced by Waddington, Needham and Brachet (quoted in Saxen and Toivonen[236]) and by Holtfreter.[146] They suggested that the neuralizing factor may be present in the competent tissue itself but inactive because it is bound or held in masked form. Induction may merely provoke the liberation of a morphogenetic substance that is present all the time in the competent tissue. Experiments by Barth[12] and by Holtfreter[144–146] provided some heavy ammunition in support of this theory. Barth[12] reported the chance observation that occasionally ectodermal explants of *Amblystoma punctata* in the absence of any notochordal or dorsal lip tissue would self-differentiate into neural plate. Holtfreter[144] had reported similar results and had pointed out that the extent of self-neuralization could be increased by treatments promoting cytolysis of some ectodermal cells in the culture. Exposure of explants to solutions of high salinity would increase cytolysis in the culture and also increase the incidence of self-neuralization in the remaining ectoderm. Holtfreter[144, 145] used competent excised ectoderm of *Triturus* and cultured it under unphysiological conditions such as high pH, a calcium-free solution, or in a hypotonic medium and noted that the frequency of self-neuralization could be considerably increased by all of these treatments.

On the basis of these results, Holtfreter[146] suggested that the inductive

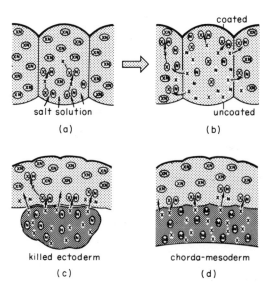

Fig. 3.11 Holtfreter's conception of the mechanism of autoneuralization and its relation to induction. The neuralizing agents X and N are present in competent ectoderm bound together as XN in inactive form. Salt solution, killed cells, cytolysing agents or chordamesoderm each can separate X and N from their complex and thus produce neuralization. (After Holtfreter,[146] taken from Saxen and Toivonen[236])

stimulus might be emerging from some of the dying cells that had cyto-
lysed in response to some of these toxic agents. Some chemical factor
released from these dying cells might subsequently reach the competent
tissue and act as an inductor.

In somewhat more sophisticated form, the idea of substances that
are universally distributed, yet have very specific effects on growth and
differentiation, is still with us. In the thirty years since the original efforts by
Spemann, embryologists have isolated innumerable growth factors and
morphogenetic substances, each specific in their effect on very selective
targets. An example is the nerve growth promoting factor (NGF) (Levi-
Montalcini and Angeletti[183]). The concentration of such substances in
the various tissues may differ but few tissues are entirely deficient.

Metabolic properties of the inductor

Attempts to identify properties of inductive tissue in terms of metabolism started with Woerdeman (quoted in Saxen and Toivonen[236]) who
found a very rapid decomposition of glycogen in cells that rolled over the
dorsal lip of the blastopore. Brachet[34] and Barth[11, 13, 14, 16] have shown
that the level of metabolism demonstrated by such indicators as oxygen
consumption, CO_2 production, and glycogen disappearance, differs in the
organizing or inducing centres compared to the surrounding tissue.
Brachet[32, 33] also demonstrated that in amphibian embryos considerable
amounts of ribonucleic acid accumulate in the lip of the blastopore and
the roof of the archenteron, and he advanced the view that ribonucleic
acid may play an important part in induction. However critically we may
view these early experiments today, they did blaze a trail and put us on
the right track, that induction must involve transmission of a chemical
message. The effect exerted on the competent tissue probably involves
an extra push improving the chances that the process of differentiation
gets off the ground. We can think of differentiation as a reaction which is
biologically possible just as chemical reactions that take place in the test
tube are thermodynamically possible. The inductor, like the catalyst or
the heat produced by the Bunsen burner, merely accelerates a reaction
which under optimum conditions could start itself.

EVIDENCE FOR THE INDUCTOR BEING A DIFFUSIBLE SUBSTANCE

Transfilter experiments

The period following the Second World War inaugurated what might
be called a second phase in the study of the organizer. The improved

biochemical methods that had become available, such as fractionation procedures, differential centrifugation, precipitation of ribonucleic acids and electrophoretic separation of proteins, were brought to bear on the problem of induction. The new generation of embryologists addressed themselves essentially to two problems: the chemical nature and identity of the inductor, and the events occurring in the reacting system as a consequence of the inductive stimulus. At the present time, we neither know the identity of the biochemical agent which presumably passes from the archenteron to the prospective neuroectoderm; nor are we certain about the reactions set into motion by the inductor on the competent tissue. We are aware that the end result is 'neuralization', but we know nothing of the intermediary processes nor of the chain of reactions that intervene to produce what we ultimately recognize as morphological or chemical differentiation.

Mangold (quoted in Saxen and Toivonen[236]) implanted into gastrulae pieces of agar which had previously been in contact with the medullary plate of an older embryo, and had noted that in the presence of such implants the overlying ectoderm began to form a neural tube. To re-investigate the problem of whether induction requires actual physical contact or whether the inductive stimulus can reach the com-

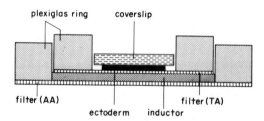

Fig. 3.12   A method for separating inductor and competent ectoderm by filters of known pore size. The inductor tissue is placed in a large filter cup and covered by the ectoderm-bearing filter. (From Saxen and Toivonen[236])

petent tissue by diffusion, Holtfreter[148] minced up inductively active archenteron tissue, placed it into a hole prepared in paraffin at the bottom of a culture dish, and covered it with cellophane. Ectoderm was transplanted on top of the cellophane membrane. Under these conditions, neural tube formation was observed. These observations are not conclusive since toxic material released by the dead tissue could have reached the ectoderm and thus stimulated autoneuralization. The results stimulated many investigators, however, to consider the possibility that the inductive stimulus can pass through membranes whose pores are sufficiently large to permit diffusion but not large enough to permit cytoplasmic contact. In the 1950s, these problems were re-investigated on a more sophisticated level by interposing transfilters of different pore size between the two components of the inductive system, namely, the archenteron roof and the neural ectoderm. The rationale of interposing filters of known pore size is simply that it permits a more intelligent guess as to the nature of the diffusing agent. At least the size of the pore diameter would provide a clue as to whether cytoplasmic processes, macromolecules or only ions can pass.

Fig. 3.13   An ectodermal explant induced through porous filter by normal tissue. Note forebrain, eye and nasal placode. (From Saxen and Toivonen[236])

Induction by conditioned medium: Niu's and Twitty's experiment

The most convincing experiment in support of the proposition that the signal sent out by the organizer is a diffusible substance was performed by Niu and Twitty.[214] Niu was alerted by the chance observation that isolated ectodermal cells that had become trapped in capillary tubes filled with coelomic fluid from spawning female newts, would transform into mesoblasts. This was a rather startling observation, which was quite out of step with traditional and more orthodox views, for mesoderm and not ectoderm is supposed to give rise to mesoblasts. How was one to explain the observation that ectodermal cells were undergoing in coelomic fluid transformation into mesodermal derivatives? Niu guessed that perhaps something had diffused into the coelomic fluid and stimulated ectodermal cells to transform into mesoblast. As so often, the difference between success and failure in science is the ability of the great scientist to recognize a gold mine, while the less endowed investigator is more likely to pass such gold mines by without recognition. If faced with a similar observation, many other scientists might have blamed their technicians, and have insisted that only they, rather than nature, could have made so outrageous a mistake as the transforming of ectoderm into mesoderm. Niu now proceeded to test the proposition that if inductor signals are biochemical substances, they should diffuse and be recoverable in a suitable fluid medium, just as the factor that can transform ectoderm into mesoblasts accumulates in the coelomic fluid. A *conditioned medium* was produced by culturing archenteron or dorsal lip tissue that was obtained from *Triturus* in modified Holtfreter solution with no proteins or amino acids present. After ten days, the inductor tissue was removed from the medium and competent ectoderm placed into the latter. After twenty-four hours of exposure to the conditioned medium,

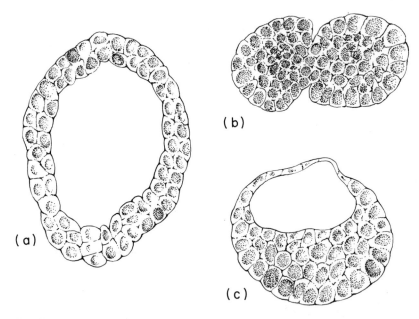

Fig. 3.14   Different cell types emerge from a sheet of ectoderm cultured in conditioned medium. A tube-like structure is shown in (**a**) and (**c**). (From Saxen and Toivonen[236])

three different cell types emerged from the ectodermal explant: nerve cells, pigment cells, and mesoblast cells. The predominance of each of the cell types derived varied with the age of the inductor tissue originally used to produce the conditioned medium. Controls consisted of ectoderm cultured in identical medium that had not previously been exposed to inductor tissue. In such 'unconditioned' medium, the ectoderm had a tendency to spread as a sheet of undifferentiated cells.

This was an important experiment, because for the first time it had been demonstrated that neural ectoderm can differentiate into nervous tissue without the physical presence of the inductor. Niu[211] interpreted his observations to mean that some substance had diffused out from the inductor into the medium, carrying the inductive message to the overlying ectoderm. Niu could further demonstrate that the age of the inductor somehow modifies the nature of the signal that is emitted. When the archenteron of seven to ten day old embryos served as the source of the inductive stimulus, nerve and pigment cells were mainly observed to grow out from the ectodermal explants whereas inductor taken from twelve to fifteen day old embryos stimulated the ectoderm to form mesoblast in 70 per cent of the cases tested. This indicates either that different inductor substances are produced with progressing age, or that a change in the original inductor molecules takes place with aging.

CHEMISTRY OF THE INDUCTOR

Evidence for RNA as the inductively active fraction

The remarkable and elegant demonstration that induction can be accomplished with diffusible substances, intensified the search for the chemical nature of the inductor. Using samples of RNA prepared from various adult tissues, Niu[212, 213] tested their inductive activity under experimental conditions. He found that RNA derived from calf kidney and thymus promoted differentiation of competent ectoderm into nerve, pigment, and mesoblast cells and, therefore, concluded that the RNA fraction of the nucleoprotein was the inductively active factor of the molecule.

Subsequently, Niu[212–213] set out to isolate the inductively active material of his conditioned medium. The solution was collected at seven to ten day intervals and subjected to various chemical tests. Fractions were tested for inductive activity and their chemical nature identified by their maximal absorption spectra in U.V. light. The absorption spectra recorded for the inductively most active substance ranged between 258 to 265 mμ. This suggested a nucleoprotein as the inductive material. In order to test whether the inductively active moiety is the protein or the nucleic acid of the molecule, the proteolytic enzyme, chymotrypsin, as well as the enzymes DNAase and RNAase were applied to the fraction isolated from the conditioned medium. According to Niu, chymotrypsin treatment did not destroy the inductive action of the isolated fraction but RNAase treatment did, a finding that led him to postulate that inductive potency must reside in the RNA of the nucleoprotein. This claim has been seriously debated and contested by other investigators.

The distribution of neuralizing and mesodermalizing factors

Different ideas about the inducing factor from those proposed by Niu were advanced by the Japanese school, represented by Hayashi[139-141] and Yamada[301-302] and the German school represented by Tiedemann[253] and the Finnish school represented by Toivonen.[254, 256] The problem of the chemical identity of the inductor is complicated by evidence presented by these investigators for the presence of a second inducing factor, the so-called mesodermalizing factor. This latter factor promotes differentiation of notochord, somites, pronephros, and tail structures.

It so happens that liver and kidney tissue are exclusively sources of factors which induce neural structure, or *neuralizing factors*.

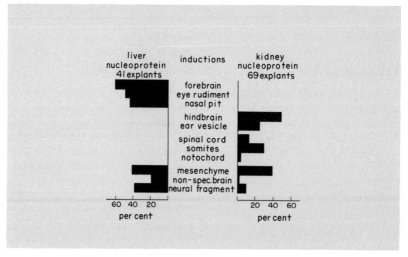

Fig. 3.15   The inductive action of nucleoprotein samples isolated from guinea pig kidney and liver tissues. (From Saxen and Toivonen[236])

On the other hand, when bone marrow is introduced into an *in vitro* system in which competent ectoderm is grown, notocord, somites and tail structures are formed instead of brain, sensory organ or spinal chord.[257] Bone marrow is therefore referred to as a source of *mesodermalizing factors*.

The neuralizing and mesodermalizing factors are probably somewhat different molecules for their dissimilarities are greater than their similarities. This conclusion is based on the observation that the two factors differ in their sensitivity to a variety of agents. For example, seeding in alkaline solution inactivates the mesodermalizing factor much more

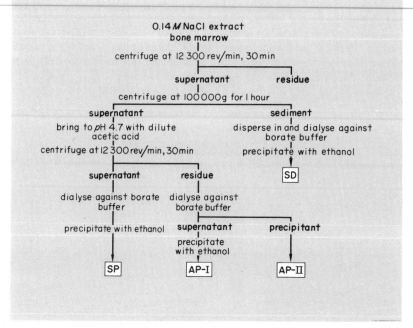

Fig. 3.16   Yamada's scheme for fractionation of the guinea pig bone marrow. (From Saxen and Toivonen[236])

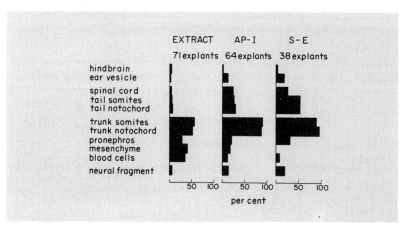

Fig. 3.17   Inductive action of extracts from guinea pig bone marrow. AP-I = acid precipitable fraction. S-E = electrophoretically separated. (From Yamada,[302] in Saxen and Toivonen[236])

rapidly than it does the neuralizing factor. Glycolic acid also inactivates the mesodermalizing factor more rapidly than it does the neural factor.

Evidence that the mesodermalizing factor resides in the intact ribonucleic protein molecule came from experiments using enzymatic degradation with either proteolytic enzymes or RNAase. Inactivation of all inductive capacity by trypsin and pepsin treatment persuaded the Japanese group strongly in favour of the protein nature of the inductor.

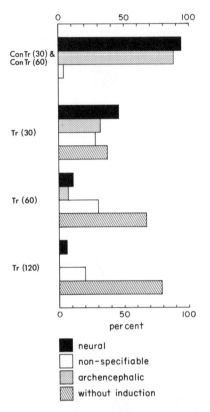

Fig. 3.18 Inductive action of liver ribonucleoprotein samples treated with trypsin for 30, 60 and 120 minutes. The control samples were incubated at pH identical with that at which enzyme-treated samples were maintained. (From Hayashi[140])

RNAase treatment did not seem to affect neural induction. Hayashi[144] did a most convincing experiment in this respect. He started with ribonucleic protein samples obtained from liver tissue and treated them with

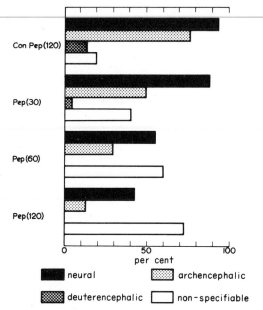

Fig. 3.19 Inductive action of liver nucleoprotein samples treated with pepsin at pH 4. Controls were incubated at pH 4 without enzyme for 120 minutes. (From Hayashi[140])

ribonuclease which removed about 99 per cent of the original RNA. Intact liver nucleoprotein samples and identical samples after treatment with RNAase showed no difference in inductive action. However

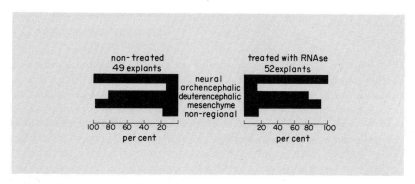

Fig. 3.20 The inductive action of purified liver nucleoprotein sample (left) and of the same sample after treatment with ribonuclease has removed more than 99 per cent of the RNA. (From Saxen and Toivonen[236])

ribonucleic acid isolated from liver and kidney, two tissues that are known to be potent inductors, totally lacked inductive activity. The situation is further complicated by the possibility that distinct and geographically separate neural inductors may exist. Liver, as was mentioned before, mainly promotes differentiation of forebrain and sensory structures and is, therefore, generally referred to as the source of *archencephalic* induction. Kidney promotes mainly spinocaudal or hindbrain development and is, therefore, considered a source of *deuterencephalic* and *spinocaudal* induction. One might interpret this to mean that specific inductors are geographically distributed along an anterior-posterior axis of the archenteron roof and that each specifies the type of structure that will differentiate in the overlying ectoderm.

The hypothesis of a cooperation between these two factors was promoted by Toivonen and Saxen[256] on the basis of the following experiments. They implanted pieces of neuralizing guinea pig liver and mesodermalizing guinea pig bone marrow into the blastocoele of amphibian embryos and noted that neither liver nor bone marrow alone produced spinal cord, but that when both of these tissues were implanted together, spinal cord was produced in a high percentage of cases.

TRANSFER OF THE INDUCTOR

Early studies: disappearance of RNA from inductor and reappearance of RNA in induced tissue

The next question to be attacked was whether there is any direct evidence for transfer of material from inductor to overlying ectoderm. The earliest proponent of transfer of material from inductor to competent tissue was Brachet.[34] He studied the distribution of pyronin-stained basophilic granules in amphibian embryos during early gastrulation.[30] The granules lose their stainability during treatment with ribonuclease, suggesting that they probably consist mainly of RNA. During invagination there is a decrease in these granules in the dorsal lip and an increase of them in the overlying ectoderm. Brachet suggested that there is probably a transfer of RNA from dorsal lip and a consequent increase in the overlying ectoderm. Rounds and Flickinger[223] measured RNA content of different fractions of homogenized dorsal lip and ectoderm at different stages of development by spectrophotometric procedures. They noted that RNA decreases in chordamesoderm during invagination just as Brachet[30] had suggested, but that ectodermal RNA does not increase simultaneously. The increase of ectodermal RNA takes place somewhat later in the neurula stage. Possibly some RNA transferred to the ectoderm is lost.

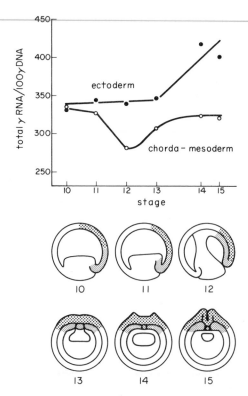

Fig. 3.21   Increase of RNA content of the ectoderm and correlated but not coinciding decrease of RNA content of the chordamesoderm during gastrulation. (From Rounds and Flickinger,[223] in Saxen and Toivonen[236])

Evidence from experiments with radioactively labelled inductors

The favourite method for demonstrating transfer of material involves labelling of the inductor tissue and then following the subsequent appearance of label in the competent tissue. In Waddington's laboratory, a series of just such experiments were carried out. Chordamesoderm of axolotl embryos was labelled with ^{14}C glycine and preferential accumulation of labelled material was subsequently noted in the nucleus of cells of the induced neural plate.[237, 275, 276] From these results the authors concluded that molecules are transferred but that massive transfer of cytoplasm does not occur. Waddington also reported that when the living organizer is replaced by killed but inductively active tissue (mouse kidney) that had previously been exposed to a label, accumulation of the label in the cytoplasm of the competent ectoderm still occurs. Preferential accumulation of label in induced tissue was also demonstrated by Ficq[86, 87] who used living organizer labelled with ^{14}C orotic acid and ^{14}C glycine and noted passage of labelled material to the reacting ectoderm. In this system radioactive material was incorporated in all those dorsal parts of the embryo that are characterized by high metabolic activity. She concluded that some macromolecules, probably protein or nucleic acid, or both, actually pass from inductor to competent ectoderm. A convincing demonstration of transfer of material was provided by the work of Rounds and Flickinger.[223] They cultured combinations of ^{14}C labelled

Type of fusion	TCA-insoluble fraction	Nucleic acid fraction
Competent ectoderm $\longrightarrow$ Ventral mesoderm—[14]C	13	15
Ectoderm—[14]C Dorsal mesoderm $\longrightarrow$	11	128
Competent ectoderm $\longrightarrow$ Dorsal mesoderm—[14]C	40	396

Fig. 3.22   Transfer of radioactive ^{14}C to originally unlabelled ectoderm cultured together with inductively active and inactive tissue labelled with [14]C. (From Rounds and Flickinger,[223] in Saxen and Toivonen[236])

inductive tissue together with competent non-labelled ectoderm. Combination of ectoderm with labelled inductively inactive ventral mesoderm served as controls. After 8 hours in culture, the two components were separated; protein and nucleic acid fractions were extracted from the ectoderm and counted for radioactivity. The highest count was recorded in the nucleic acid fractions obtained from competent ectoderm cultured for 8 hours with labelled inductively active dorsal mesoderm.

Separation and recombination of the components of tissue interacting systems: Grobstein's experiments

It is obvious that the study of transfer during induction could be much better analysed in a system where it is possible to separate the components of the tissue interacting system and to recombine them at will. Grobstein's[103] experiments designed to analyse the interaction of *nephrogenic mesenchyme* and spinal cord to form kidney tubules met these criteria superbly. In Grobstein's system, nephrogenic mesenchyme, i.e. that portion of the intermediate mesoderm normally forming kidney tubules of nine-day-old mouse embryos, can be induced to form kidney tubules *in vitro* if grown in proximity to spinal cord. Morphologically, initial tubule formation is discernible after 24 hours of culturing. Spinal cord alone, from a variety of tissues that have been tested, is capable of inducing nephrogenic mesenchyme to form kidney tubules. Removal of the spinal cord before completion of 24 hours in culture will effectively prevent differentiation of tubules. Removal of the spinal cord after 30 hours of culturing will not interfere with kidney differentiation. If the cord is removed at or before 30 hours of culturing the tubules already formed will persist though somewhat smaller in size. Mobility of the inductive material from dorsal spinal cord to nephrogenic mesenchyme has been demonstrated by Grobstein[109, 113] in a series of experiments. Dorsal spinal cord of an 11-day-old mouse was placed in an assembly separated into two compartments by a filter. On one side of the filter, into a medium containing tritiated leucine, was placed embryonic spinal cord. Onto the filter was placed nephrogenic mouse mesenchyme. After 2 hours the mesenchyme was removed and radioactivity of the material deposited on the filter and into the overlying mesenchyme was determined.

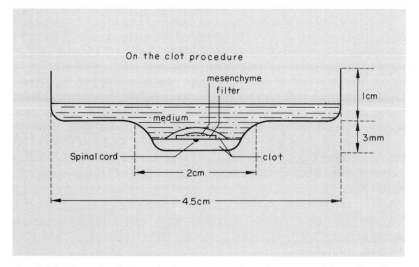

Fig. 3.23   Dorsal spinal cord of a mouse embryo is separated by a filter from nephrogenic mesenchyme.

When the intervening membrane had an average pore diameter of 0·5 μm, interaction was at maximum intensity up to 30 microns. It declined from 60–80 μm and was absent at distances beyond that magnitude. Membranes whose average pore size is 0·7 μm limited the distance of effective interaction. Cellophane membranes that were interposed between the two tissue components cut off the interaction altogether, but when a small opening was made through the membrane interaction occurred through the hole and the response was localized immediately over it.

The latent period of 30 hours may constitute the time necessary for the production of inductive material by the inducing spinal cord. It may also constitute the time required to accomplish transit of inductive material from cord to competent mesenchyme or it may be the time necessary for the reacting tissue to respond to the molecules deposited by the inductor on its surface. Unfortunately there is a scarcity of similar experiments on the dorsal lip-neural plate interaction. A notable exception is the study of Saxen[235] who used a modification of the culture system devised by Grobstein, and reported that cultures of competent ectoderm separated from heterogenous inductor tissue of *Triturus vulgaris* by a filter of pore size of 0·8 μm did neuralize in 50 per cent of the cases tested.

Intracellular localization of the inductor

The question of where the inducing factor is localized and to what cell organelle it may be bound was first attacked by Brachet. Pellets of different cell components obtained by ultracentrifugation of liver and kidney cells were implanted into amphibian blastocoeles and tested for frequency of neural induction that was obtained. The most active fraction turned out to be the microsomes which in amphibian embryos induced archencephalic as well as spinocaudal structures in high proportion. The yolk fraction was the weakest and least active in neural inductive capacity. Localization of inducing factor on microsomes does not, however, in itself imply that whole microsomes or whole ribosomes are actually transferred from the archenteron roof to the ectoderm in the process of induction.

SOME SPECULATIONS ABOUT THE DISTRIBUTION OF INDUCTIVE SUBSTANCES AND THE MECHANISM OF INDUCTION

The results of the numerous experiments performed to identify the chemical nature of the inductor strongly suggest that embryonic inductors may be widely distributed in the biological household. Consider the following observation. Isolated gastrula ectoderm by itself has no inductive capacity. After treatment with ethanol, it can promote the construction of forebrain, if placed next to competent ectoderm. Treatment with phenol will convert it into an inductor of hindbrain, ear vesicles, and in some cases, spinal cord. The extract of trunks of 9-day-old chick embryos centrifuged for 2 hours at 105 000 g has only weak inductive capacity. The same supernatant heated at 95°C before precipitation with ethanol induces forebrain and eyes. These experiments can be interpreted to mean that the inductor may be universally distributed in all tissues but may be tied to an inhibitor which may be released by these treatments. The natural inductors that operate *in vivo*, may be merely those that exist already in free form, and are released from the inhibitors during gastrulation without help and assistance of the experimenters.

Recently it has been proposed that the mechanism of induction may operate by selective destruction of the cells of competent tissue. These conclusions have been advanced by Ave, Kawakami and Sameshima[9] on the basis of the following observations. Electron microscopic study of the presumptive gastrula epidermis by these investigators revealed that the gastrula ectoderm consists of two types of cells, i.e. electron-dense cells and transparent cells. Mesodermalizing treatment seems to result in the destruction of the electron-dense cells, while neuralizing treatment is followed by the destruction of the transparent cells and selective survival of the dense cells. Electrophoretically dissociated cells of the undetermined ectoderm of the gastrula could be distinguished into at least three zones. The cell number in these zones in a cathode to anode direction corresponds to 35, 41 and 24 per cent respectively. Treatment of gastrula epidermis with bone marrow extract, a tested mesodermalizing agent, promoted survival of about 66 per cent of the original population of the dissociated presumptive epidermis, all of which distributed at the anodal zone. Conversely, when dissociated cells of the presumptive epidermis were subjected to neuralizing treatment, approximately 38 per cent of the original population was recovered, all of which distributed at the cathode.

These results commend the hypothesis that perhaps the presumptive epidermis prior to the inductive action of the archenteron is not as homogenous a cell population as heretofore believed, and furthermore that one of the effects exerted by the archenteron on the competent ectoderm may consist in the selective killing of one of the two species of cells, already present, long before inductive interactions take place.

4

Models of Tissue Interaction

SOME EARLY INSIGHTS REGARDING THE MECHANISMS OF INDUCTION

Speculations and generalizations

In the absence of any certain knowledge about the intermediary processes taking place between the elaboration of the inductor and the final product, 'differentiation' of competent tissue, there is no shortage of speculation (Tiedemann,[252, 253] Wolff,[299, 300] Yamada[303]). Most of these speculations are merely restatements of observations expressed in a more sophisticated manner. The majority of experimental observations accumulated over the last 30 years have led many investigators to suggest that the inductor is a sort of signal alerting the immature cell to realize its potential.

Most likely, the inductor is not just another set of instructions but rather a stimulus that activates 'competent' embryonic cells that have already received their own genetic assignments to continue on the path of differentiation. The concept of 'competence' in this context means nothing more than the observation that the genes of the reacting tissue will generally determine the end product of their differentiation. The initiation of all those events leading to differentiation require some sort of a cue; which seems to be provided by some substance transferred from the neighbouring tissue to the competent cells.

Genotype and inductor: Schotté's experiment

A classic experiment by Spemann and Schotté[245] demonstrated in a very clever manner one kind of relationship between inductor and competent tissue. In amphibian embryos, ectodermal mouth parts are induced by the oral endoderm. Some urodeles, notably *Triton* larvae, normally develop ventrolaterally to the eye a pair of long club-shaped filaments called balancers. The tadpoles of frogs have no such balancers but do have mucus-secreting glands or suckers, which urodeles lack. Spemann and Schotté[245] transplanted ectoderm from an early frog gastrula to the head area of a newt embryo, in such a way that the grafted ectoderm came to lie in the future mouth region. The frog skin thus transplanted to the newt was induced to form a head armed with typical frog suckers. In the reciprocal experiment, Spemann and Schotté[245] transplanted flank ectoderm from newt embryos to the mouth region of frogs. The mouth structures developing were typical of the newt donor. These results were interpreted to mean that apparently the embryonic frog ectoderm responded to the inductors of the newt head region by forming mouth structures typical of frogs, i.e., having suckers. Conversely newt ectoderm responded to the inductive message of frog oral endoderm by forming mouth parts typical of the newt genotype.

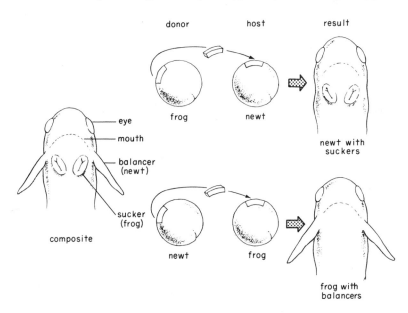

Fig. 4.1   Schematic representation of Spemann and Schotté's experiment. Note prospective oral epithelium of a frog transplanted to a newt gave rise to oral suckers, typical of the mouth region of a frog. Conversely oral epithelium from a newt transplanted to a host frog gave rise to balancers, typical mouth structures found in newts. (After Barth[17])

It is as if competent ectoderm is pushed by the inductor to start its construction job, the details of which are provided, however, by its own genes. One way of thinking about the inductor is along the following analogy. Suppose we compare embryonic cells to a group of students sitting in a row. Each one is given a specific task on the first school-day; student No. 1 is to wash the blackboard, student No. 2 is to open the window, student No. 3 is to lay out the chalk, student No. 4 is to arrange the papers and student No. 5 is to set up the projector. When the teacher enters the classroom and barks out a command each of the five participants gets up and carries out his highly specific assignment in response to the same general command.

Time requirement of induction

Another hint that the inductive stimulus may be relatively unspecific comes also from the observation of the minimum time required for induction to take place. Johnen[162] reported that when she combined isolated gastrula ectoderm with archenteron *in vitro* the latter had to be present for only five minutes in order to promote formation of brain and eye structures. She suggested that during this brief period, the message of the inducing agent is either recorded and perhaps amplified to switch the gears of cell differentiation into a new direction, or the inductor triggers the release of a mechanism that is all set to go.

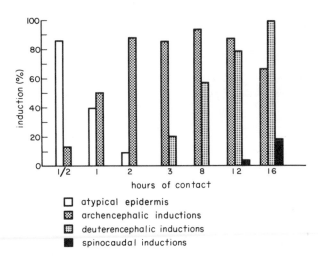

Fig. 4.2 Percentage of archencephalic, deuterencephalic and spinocaudal inductions when archenteron roof was used as inductor. The time of contact varied from ½ to 16 hours (From Saxen and Toivonen[236])

The search for simple interacting systems

Induction of the neural tube from competent ectoderm turned out to be a rather difficult system for the analysis of the mechanisms and events that are initiated in response to the inductive stimulus. With the refinement of methods of investigation that have become available since the end of World War II, a number of progressive embryologists have turned their attention to other, simpler systems and tried to identify in biochemical terms just what takes place when a population of embryonic cells change course and differentiate under the impact of the inductor stimulus. In their efforts to elucidate the mechanisms that operate in tissue interaction, they focused their attention hopefully on such interacting systems as 'Wolffian lens regeneration' (Yamada[305]), 'epithelial mesenchyme' interaction (Grobstein[101–105, 108, 111, 112]), cartilage induction (Holtfreter,[149] Holtzer,[151–155]) and limb induction (Amprino,[1] Goetinck,[96] Saunders,[231, 233, 234] Zwilling[310–314]).

WOLFFIAN LENS REGENERATION

Lens regeneration from the iris

The regeneration of the lens from the iris has been favourite material for study. In many species of salamanders, a new lens will form from the iris after lens extirpation. Apparently there is a factor present in the optic chamber that is necessary for the initiation of the transformation of iris into lens.[68] When an isolated iris is cultured *in vitro* no lens regeneration takes place but if the iris is transferred to the chamber of the eye the iris will transform and regenerate the missing lens. On the other hand, if the whole eyeball is cultured without a lens the iris will regenerate the missing lens. This suggests that some inductive substance emanates from the optic chamber, which can cause the iris to lose its differentiated state and transform into another tissue. (For review see Reyer,[219] Stone,[248] Yamada[305]) The factor inducing lens regeneration probably comes from the retina, for if the neural retina is removed from the eye along with the lens, regeneration of the lens from the iris does not start until the neural retina begins its regeneration. Repression of neural retinal regeneration by removing pigmented retina and choroid together with neural retina is followed by failure of lens regeneration from the iris.[248] Although the transformation of iris into lens is not strictly identical with induction of undifferentiated tissues, this process can perhaps provide a model for the changes that are initiated in competent tissue under the influence of inductors.

Cellular and biochemical changes during Wolffian regeneration

Some of the changes in the iris that precede the emergence of regenerated lens from the iris have been followed by Yamada.[305] The basic cellular changes are the following:

1. A *depigmentation phase*, lasting from 3–5 days after the removal of the lens during which the iris cells gradually discharge their pigments, starts the sequence of events. Radioautographs have revealed that during this phase the cells of the iris, which under normal conditions are non-dividing and consequently show only small amounts of ^{3}H thymidine uptake, are stimulated to engage in very active DNA synthesis, presumably in preparation for division.

2. The second phase, or *multiplication phase*, is a period of active mitosis of the depigmented cells. At that stage the DNA synthesis which precedes cell division has come to a standstill and the re-appearance of nucleoli implies that ribosomal synthesis is probably reaching its peak. At any rate, soon after removal of the lens the incorporation first of tritiated thymidine and subsequently of tritiated uracil by cells of the regenerating iris is considerably enhanced.

3. The third phase or the *fibre differentiation phase* shows histologically visible changes such as elongation of the daughter cells that are the product of the preceding mitoses, and the disappearance of nuclei immediately after entering the stage of fibre differentiation. The daughter cells next begin to show three of the lens specific proteins: the α, β, and γ crystallines. The α and β crystallines appear first, the γ appears somewhat later.

If the regenerating iris is removed to a site away from the influence of the optic cup (such as the brain) the transformation from iris to lens will not proceed to its final endpoint. In such an unsuitable environment, the regenerating iris cells that have started the process of depigmentation can continue to proliferate but will fail to differentiate into fibre cells. Only just at the emergence of the lens proteins can the iris be removed from the optic cup environment and still continue further differentiation independently on its own. It would appear from these observations that the cells of the lens require the influence of the optic cup until the time when the specific lens proteins have been synthesized in the differentiating fibre cells.

Hypothetical scheme of gene activation during Wolffian regeneration

Perhaps the inductor responsible for lens regeneration may influence the pattern of gene activation in the regenerating iris. The earliest changes detected in the iris after lens removal are indeed an increase in the incorporation of uracil and cytosine into the nuclear fraction. All subsequent cellular events that have been observed in iris-lens regeneration may be secondary consequences of this primary effect. The massive increase in RNA that has been noted in regenerating iris cells, although probably of the ribosomal type, may merely accompany the simultaneous initiation of some specific messenger RNA molecules. (The sudden appearance of such new messenger molecules would certainly be masked by the simultaneously occurring massive emergence of ribosomal RNA). Inhibition of lens regeneration by actinomycin D points to the possibility that under the influence of an inductive stimulus, the pattern of gene activity of regenerating iris cells is switched to cause differentiation into another type.

MECHANISMS INVOLVED IN INDUCTION OF EPITHELIAL STRUCTURES

Discovery of epithelial–mesenchymal interaction

In 1953, Grobstein published a series of papers that have greatly influenced our outlook on inductive tissue interactions. He reported a series of experiments in which the rudiments of the submandibular gland of 13-day-old mouse embryos were separated with 3 per cent trypsin solution and the two components of the rudimentary glands were then grown in culture and their differentiation observed.

The two components of the salivary primordium, the epithelium and the mesenchyme, can be easily separated. Both can be grown in culture separately and remain viable. Under good culture conditions, the mesenchyme capsule produces freely moving mesenchymal cells, while the epithelial bud loses its shape and transforms into a flat sheet of cells. When the two components are first separated and then recombined and placed near to each other, the mesenchymal cells will again come to surround the epithelium and the submandibular epithelium will begin to sprout tubules and take on the appearance of an exocrine gland. (For a review see Grobstein[103]) Grobstein's experiment stimulated several investigators, including Auerbach[3] and Wessells,[285-290] to address themselves to the problem of the inductive process in a variety of other epithelial mesenchymal interactions. Auerbach[3] demonstrated dependence on mesenchymal induction for differentiation of the thymus epithelium; and Wessells[286-288] for pancreatic differentiation. *Epithelium-mesenchymal interactions* have been shown to be required for differentiation of a large number of organs since the original study of Grobstein. In a great many cases the mesenchymal capsule exerts an inductive action that is necessary for the differentiation of its associated epithelial rudiment. (For review see Wolff.[300])

Specificity of mesenchymal induction

To test the specificity of the effect emanating from the mesenchyme, submandibular epithelium was cultured in the presence of mesenchyme derived from nonsalivary gland organs such as somites, lung, heart and lateral plate etc. With foreign mesenchyme, the submaxillary epithelium always failed to differentiate into tubular structures.[102] Apparently, the inductive stimulus for normal differentiation of submaxillary epithelium is specific and can be provided only by the capsular mesenchyme of the submaxillary gland itself. The high specificity of the mesenchymal effect demonstrated in the submaxillary gland is not typical of all mesenchyme-epithelial interactions however. *In vitro* epithelium of pancreatic rudiments obtained from 11-day-old mouse embryos, just like salivary epithelium, can differentiate only if grown in combination with mesenchyme, but the requirement can be met by mesenchyme derived from many other sources.[288]

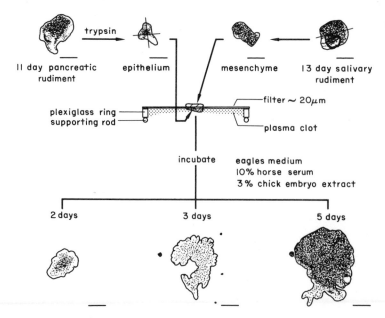

Fig. 4.3 Schematic representation of experimental procedure for pancreatic morphogenesis *in vitro*. The appearance of typical pancreatic epithelia after various periods of culture is indicated at bottom. (From Rutter, Wessels and Grobstein[230])

The critical period of induction

Thirty hours of actual mesenchymal-epithelial interaction are required to bring about changes in the epithelium that will become recognizable subsequently as pancreatic differentiation. After 30 hours the mesenchyme can be removed from the culture but it takes an additional 18 hours for pancreatic differentiation to proceed to a point where histological differentiation is seen. The only clue to what may be happening during this critical 18-hour period comes from the observation that differentiation of pancreatic epithelium even after exposure to the inducing mesenchyme for 30 hours can be prevented if the epithelial mass is fragmented or spread too thin.[230, 288] If fragmentation is delayed until 48 hours after exposure to mesenchyme then this disturbance no longer hinders pancreatic differentiation.

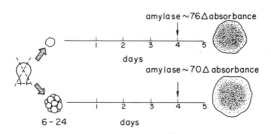

Fig. 4.4   Comparison of rates of differentiation in pancreas epithelia of differing initial mass. When a quarter-sized piece of 11-day epithelium is compared with up to 24 pieces clustered together, opacity due to zymogen granules and rise in amylase activity appear at the same time. (From Wessels[288])

Seventy-two hours after culturing, the prospective pancreas cells have reached the stage where they are no longer sensitive to actinomycin D, as shown by the observation of Rutter, Wessells and Grobstein,[230] who have demonstrated that zymogen synthesis can be initiated and continued in the presence of this drug at a concentration which would have effectively inhibited pancreatic differentiation and blocked zymogen synthesis in younger pancreatic precursor cells. Although between the 30th and 40th hour of culturing, the prospective pancreatic epithelium shows no signs of histological differentiation, a critical change must occur during this 'covert' phase that precedes the 'overt' phase when zymogen granules and acinar formation begin to become visible. The changes taking place during this 'covert' phase in the competent epithelium have not yet been identified, but it seems likely that by focusing attention on the events taking place in the pancreatic epithelium during this restricted period of differentiation, information may be

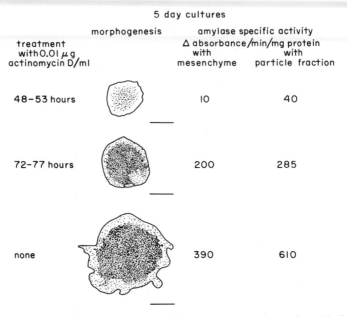

	5 day cultures		
treatment with 0.01 μg actinomycin D/ml	morphogenesis	amylase specific activity Δ absorbance/min/mg protein with mesenchyme	with particle fraction
48–53 hours		10	40
72–77 hours		200	285
none		390	610

Fig. 4.5   Effect of actinomycin D on differentiation of pancreatic epithelium. Cultures were treated with actinomycin D for a 5-hour period at various times of culture. The cultures were then incubated for the remainder of the 5-day period. Cultures treated between 48 and 53 hours grow, but amylase activity is depressed. Treatment at 72 hours gives cultures whose amylase activity and zymogen granule concentration approaches that of normal cultures. (From Rutter, Wessels and Grobstein[230])

gained about the reaction released by the inductor in the responding cell. To quote Rutter, Wessels and Grobstein[230] 'The experimental system [i.e., the rudiment of the pancreas] appears to be willing to tell us more.'

The distribution and suggested mode of action of the mesen-chymal inductor

Pancreatic differentiation can proceed *in vitro* even if the mesenchyme is separated from the epithelium by a 20 μm pore size millipore filter. Apparently, therefore, in this system a macromolecular substance may be transmitted from inductor to competent epithelium. The finding that the requirement for pancreatic differentiation could be met by mesenchyme from any source led Rutter, Wessells and Grobstein[230] to testing for the presence of this inductive agent in chick embryo extracts. It was found that raising the concentration of chick embryo extract in the culture medium from the usual 3 per cent to 20 per cent promoted differentiation of the pancreatic epithelium into typical gland structures in the complete absence of mesenchyme. This implies that this inductor may be widely distributed and not limited to the mesenchyme.

Table 4.1   Requirements for *in vitro* differentiation of pancreatic epithelium. From Rutter, Wessels and Grobstein.[230]

Experi- ment	\ Mesen- chyme	Culture conditions \ Chicken embryo* extract (%)	Morpho- genesis†	Analysis 5-day cultures \ Amylase specific activity (Δ Absorbence/min/mg protein) Mesenchyme Epithelium	
1	+	0	–	None detected (<0·2)	4
2	+	3	+ +		250
3	+	20	+ +		220
4	–	3	–	None detected (<0·5)	
5	–	10	+ +		165
6	–	20	+ +		240

* Concentration based on volumes of crude embryo extract [embryos extracted with an equal volume of Tyrode's solution under defined conditions].
† Microscopic evidence of zymogen granules in cultures.

They suggested that perhaps a long-lived messenger RNA molecule necessary for the synthesis of a tissue specific pancreatic enzyme protein like amylase may be produced between 48 and 72 hours after contact with the inductor. The mesenchymal inducer may merely switch on the synthesis of the required messenger molecules. (For review see Rutter *et al.*[230])

Passage of mesenchymal inductor to competent epithelium

The first clue that mesenchymal tissue may actually elaborate a product that passes to and accumulates at the surface of the reacting epithelium came from the work of Kallman and Grobstein.[163–165] These investigators found that during interaction of pancreatic or salivary epithelium with mesenchyme, collagen fibres accumulate at the surface of the epithelium that faces the mesenchyme. In order to test whether these collagen fibres are actually transferred from the mesenchyme to the reacting epithelium, one of the two tissues was labelled with tritiated proline and cultured in an assembly with a millipore filter between them.

In cultures containing mesenchyme which was previously exposed to tritiated proline, the collagen subsequently appearing at the surface of the epithelium was labelled. In cultures containing epithelium fed radioactive proline, the collagen that was subsequently deposited at the epithelial surface remained unlabelled. These data were interpreted to mean that the collagen deposits that appear on the surface of the competent epithelium are actually derived from the mesenchyme. The importance of the initial collagen deposition for epithelial morphogenesis is brought out by experiments in which application of the enzyme collagenase to differentiating salivary epithelium resulted in the destruction of newly formed collagen. The removal of this newly deposited embryonic collagen from salivary epithelium by exposure to collagenase caused the epithelium to remain a smoothly contoured nonglandular tissue. It is therefore reasonable to suppose that the deposition of collagen on the epithelial interface might constitute one of the steps in a series of sequential interactions involved in the 'induction' of pancreatic and salivary epithelial structures.

CARTILAGE INDUCTION

The chemical profile of cartilage

Another favourite system for study of the events following induction is provided by the differentiation of vertebrate cartilage. In biochemical terms a cartilage cell is relatively simple and can be defined as a cell that has acquired the ability and the equipment necessary to synthesize, besides the many enzymes and other proteins it shares with all cells, two specific molecules, chondroitin and keratosulphate. Obviously therefore, the biochemical profile of cartilage is much more easily identifiable than that of the nervous system or probably of any other highly specialized tissue. Of the specific macromolecules found in cartilage, chondroitin sulphate lends itself best to assay procedures and is specific enough to characterize cartilage.

Induction of mesodermal somites to form vertebral cartilage

The system which has been analysed by Holtzer[151–155] and by Lash[174–177] and Lash, Holtzer and Whitehouse[180] concerns the differentiation of vertebral cartilage. The results of their earlier work clearly demonstrated that an inductive stimulus is required for the differentiation of mesodermal somites into cartilage. The story of cartilage differentiation was started by Holtzer and Detweiler[155] who first showed that in amphibian embryos extirpation of the spinal cord during development

prevents the chondrification of the somite material destined to form
vertebrae. Salamander embryos that have their spinal cord removed fail
to form vertebral cartilages. But somite material, in the presence of either
ventral spinal cord or notochord, is induced to form vertebral cartilage.

Dependence of somite mesoderm on spinal cord and notochord for
cartilage differentiation can also be demonstrated *in vitro*. Chick somites

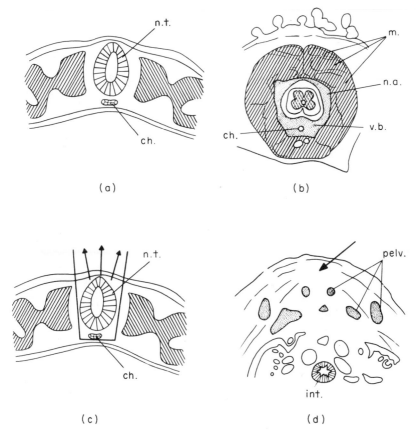

Fig. 4.6   The inducing role of chorda and spinal cord in the formation of the
vertebrae. **(a)** and **(b)** Cross-section of a normal chick embryo after 40 hours and
7 days of incubation. **(c)** Removal of the chorda and spinal cord at 40 hours. **(d)**
The resulting effect at 7 days: no vertebrae forming. The arrow indicates the site
they should occupy in the normal embryo. n.a., neural arch; ch., chorda; v.b.,
vertebral body; int., intestine; m., muscles; pelv., pelvic girdle; n.t., neural tube.
(From Strudel quoted in Wolff[299])

grown *in vitro* will differentiate into cartilage only if cultured together with ventral spinal cord or with notochord of mouse or chick. No other tissue has been found to possess similar inducing ability. Somites grown *in vitro* in the presence of anterior spinal cord or notochord will transform into cartilage on the fourth day of culture, but the presence of the inducing tissue is not required continuously to assure cartilage differentiation. The inducing tissue can be withdrawn after a few hours of culturing without impairing chondrification. This indicates that the stimulus exerted by the

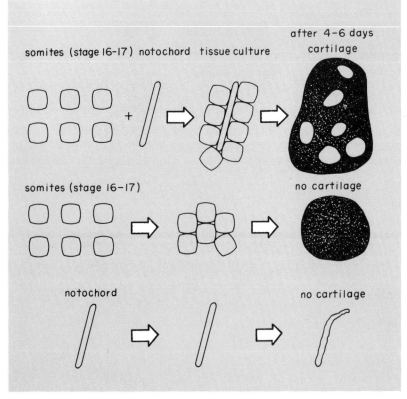

Fig. 4.7 Schematized interaction between somites (stage 16–17) and notochord from 60-hour-old chick embryo and cartilage formation *in vitro*. (From Zilliken[307])

inductor probably triggers only a single reaction, requiring only a few hours for its completion. Once the critical reaction has been elicited in the competent tissue, all the subsequent additional steps leading to the formation of visible cartilage can be carried out without the continued presence of the inductor. The experiments on which these conclusions are based have been described by Lash,[174, 176, 177] and Strudel.[249]

Chemical events intervening between the onset of induction and completion of differentiation

Acting on the belief common to many biologists that chemical differentiation precedes histological differentiation, Lash,[174–177] Lash *et al.*[178, 180, 181] and Marzullo and Lash[190, 191] tested the proposition that one of the events generated by the inductor may be the formation of tissue-specific molecules that are uniquely and exclusively present in cartilage. Chondroitin sulphate meets this criterion superbly for, as already explained, it is a characteristic if not unique product of cartilage.

In order to determine whether presence of inductor influences the sequence of biochemical reactions leading to the formation of chondroitin sulphate, Lash[174] and Lash, Holtzer and Whitehouse[180] cultured mouse somites with and without inductor in medium containing radioactive Na_2SO_4. At intervals the cultures were fixed and prepared for radioautography. The radioautographs revealed that initially the inducing tissues (cord or notochord) incorporate sulphate just as well as the somites. The incorporation of sulphate by somites was considerably enhanced however when somites were grown for 4 hours in contact with either notochord or spinal cord as judged by increases in grain counts in the radioautographs of somites.

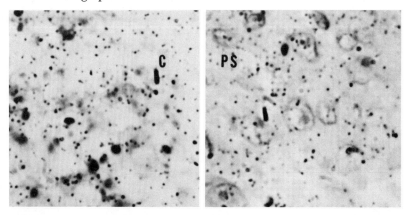

Fig. 4.8 Radioautographs of cultures of somites grown for 6 hours in medium containing radioactive Na_2SO_4. On the right, culture of somites from which polysaccharides (PS) had been extracted with 30 per cent KCl and 2 per cent K_2CO_3 for 72 hours. On the left, section of control tissue (C) extracted with water for 72 hours. Extraction for polysaccharides did not diminish components responsible for sulphate incorporation, as judged by grain counts. (After Lash[174])

In a subsequent experiment Lash[174] extracted polysaccharides from the cultures grown in medium containing radioactive Na_2SO_4 prior to preparing radioautographs. Radioautographs of cultures of somites from which polysaccharides had been extracted revealed no diminution in grain counts. These findings suggested to Lash[174] that one of the metabolic events stimulated in the somitic mesoderm by the inductor is the fixation of sulphate in preparation for the synthesis of chondroitin sulphate, but that initial sulphate fixation is not in the form of polysaccharides. To determine the manner in which early sulphate fixation occurred, cultures of spinal cord and somitic mesoderm grown on medium containing radioactive Na_2SO_4 were treated with RNAase prior to fixation and prepared for autoradiography. Autoradiographs of

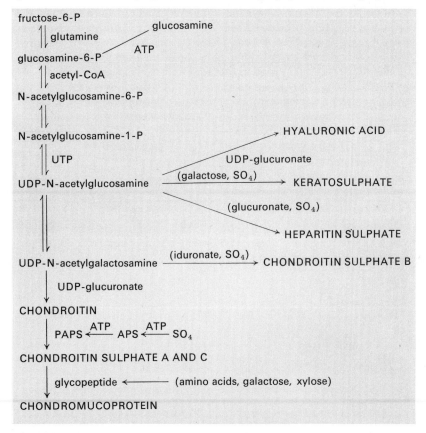

Fig. 4.9   Biosynthetic steps in the formation of UDP-N-acetylgalactosamine, the direct precursor (with UDP-glucoronate) of chondroitin sulphate. (After Lash[176])

such cultures revealed a marked diminution of grain counts. This suggested to Lash[174] the probability that the initial sulphate is fixed as a ribonucleoprotein sulphate, which is lost to the cell upon destruction by the enzyme RNAase.

This initial step in sulphate binding must occur prior to visible differentiation. The sulphation of acid polysaccharides is effected by the transfer of sulphates from active sulphate compounds to suitable acceptor molecules. The sulphation of chondroitin sulphate or kerato sulphate, two of the tissue-specific molecules of cartilage, requires enzymes transferring sulphate from active sulphate compounds such as 3-phosphoadenosine-5-phosphosulphate (PAPS) to other molecules along the following pathway.

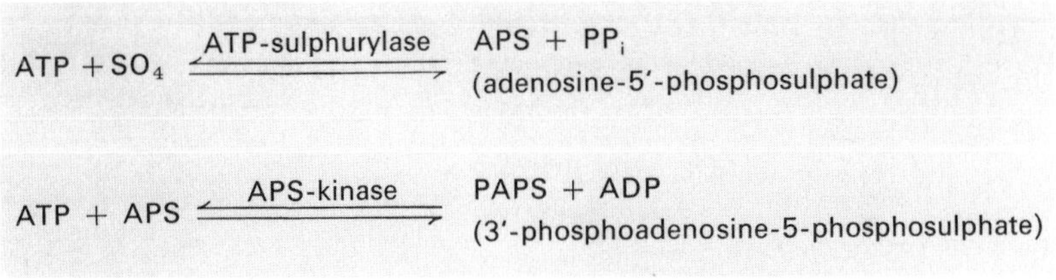

$$ATP + SO_4 \xrightarrow{\text{ATP-sulphurylase}} \begin{array}{l} APS + PP_i \\ \text{(adenosine-5'-phosphosulphate)} \end{array}$$

$$ATP + APS \xrightarrow{\text{APS-kinase}} \begin{array}{l} PAPS + ADP \\ \text{(3'-phosphoadenosine-5-phosphosulphate)} \end{array}$$

Fig. 4.10   Scheme for the formation of PAPS (After Lash[176])

It would appear from Lash's experiments, therefore, that the inductor influences not the last step in the sequence of reactions leading to chondrogenesis but more likely an earlier one, perhaps the synthesis of active sulphating compounds like PAPS. This makes sense if one remembers that the inductive stimulus can be withdrawn 4 hours after placing somites in culture but that the end effect, namely, cartilage formation, does not become visible nor does chondroitin sulphate become detectable until 3 or 4 days of culturing in an *in vitro* system. Obviously the somitic mesoderm can complete all the steps leading to the visible appearance of chondrogenesis after the inductor is withdrawn. Unfortunately, the immediate changes brought on in the mesoderm during this critical period of induction remain invisible to the microscopist and defy resolution by present biochemical procedures.

Metabolic response of somites to inducer

The possibility that the synthesis of an active precursor molecule like PAPS may be influenced by notochord or spinal cord induction was next investigated (Lash,[176] Zilliken[307]). It turned out that embryonic somites, if cultured alone, will not synthesize PAPS but somites cultured in the presence of the notochord or spinal cord will start synthesizing PAPS on the second day of culture. Since PAPS does not appear in somites until the second day of culturing, whereas withdrawal of the inductor after 4 hours does not stop subsequent cartilage synthesis, the appearance of PAPS must be a second or third order consequence of inductive events.

The idea that during the induction of somites entirely new metabolic pathways are elicited that are not operating at all before the inductor exerts its effect has had to be revised since it was first proposed. Recent experiments by Lash[176-177] have shown that the 50-hour chick embryo is capable of a certain amount of low level chondroitin sulphate synthesis. *In vivo* the 50-hour chick embryo has the ability not only to synthesize low levels of chondroitin sulphate but also to metabolize

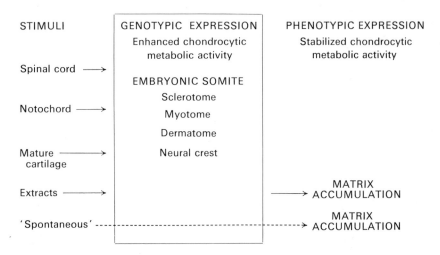

Fig. 4.11   Diagram of cartilage promoters and their effect upon expression of the somite's chondrogenic bias. According to Lash, embryonic somites have a strong chondrogenic bias before they are induced *in vitro*. Only after appropriate stimulation does the pattern become stabilized and matrix accumulate. (From Lash[177])

glucose amines and synthesize PAPS. The latter compound can actually be extracted from the *in vivo* embryo. Lash[176] suggests that the reason that in culture, pre-induced somites will not ordinarily synthesize this compound may be because during the procedure preparatory to culturing and explanting, the required enzymes may have been destroyed and cannot be reassembled except with the aid of an inductor. By the time the chick embryo has reached the 10-day stage, the ability to synthesize PAPS and sulphated polysaccharides becomes restricted exclusively to chondrogenic tissue and is completely absent in muscle, kidney, retina and liver. Subject to confirmation of these observations, one may speculate that the inductor does not stimulate the opening up of new pathways leading to differentiation, but may merely stabilize selected metabolic patterns and allow the accumulation of end products. Marzullo and Lash[190-191] and Lash[177] have since suggested that the induction exerted by notochord and spinal cord on somite cells may merely serve to stabilize an already existing biochemical reaction by blocking alternate pathways, thus leading to the accumulation of selected end products, such as chondroitin sulphate in cells destined to become cartilage. Figure 4.9 shows some of the pathways that may be blocked by the inducer and, by implication, those that may be enhanced as the result of closing all biochemical avenues other than those leading to chondroitin sulphate synthesis. According to this model, induction may not involve the acquisition of a new pattern but rather the selective enhancement and stabilization of an already existing one.

GENETIC CONTROL OF INDUCTIVE INTERACTION

Statement of problem

It is generally assumed that inductors act by repressing or derepressing genes during development, and in this way initiate the sequence of processes commonly referred to as 'differentiation'. The reciprocal assumption, namely, the possibility that the mechanism of inductive tissue interaction is in turn influenced by certain genes, has been less extensively investigated.

The increasing number of gene mutations affecting limb development that have been identified in the last decade, have stimulated both geneticists and embryologists to concentrate their search on some gene-controlled inductive processes underlying limb morphogenesis.

The discovery of the apical ectodermal ridge and its role in limb morphogenesis

In 1948, a young embryologist, John Saunders, reported a very interesting set of observations. Studying limb development in chick embryos, he noted that the ectoderm over the limb bud formed a rather conspicuous ridge,[231] which he referred to as the *apical ectodermal ridge*, or *AER*.

When Saunders surgically removed this ectodermal ridge from the mesenchyme of the limb bud in a 3-day-old chick embryo, the distal parts of the wing failed to form.

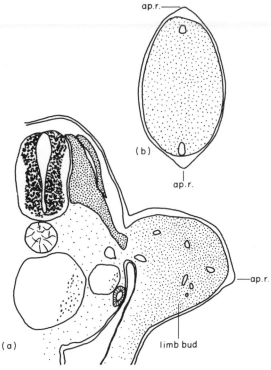

Fig. 4.12   (a) Advanced limb bud with apical ridge (ap.r.) in chick embryo and (b) cross-section from forelimb bud in rat. (From Balinsky[10])

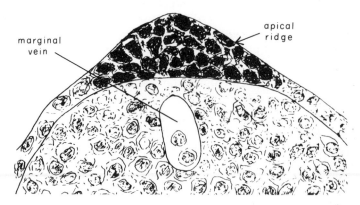

Fig. 4.13   Cross-section of the limb bud in a rat showing apical ridge stained for alkaline phosphatase. (From Balinsky[10])

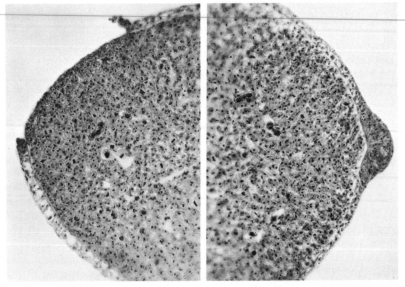

Fig. 4.14 Apex of wing bud from which the epidermis has been removed (left) and control normal wing bud with apical ridge (right). (From Saunders[231])

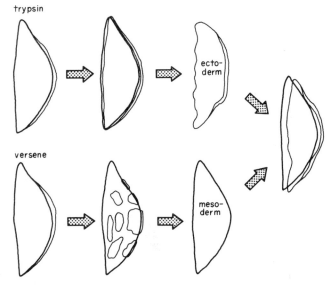

Fig. 4.15 Chemical separation of the limb bud mesoderm from epidermis by trypsin and versene digestion and subsequent recombination. (From Balinsky[10])

This suggested to him that perhaps the apical ectoderm might act as an inductor of limb differentiation. Subsequently, Zwilling[313] added a further refinement. He treated the limb buds with trypsin or versene and was then able to peel off the ectoderm of the limb bud from the underlying mesenchyme.

Applying this method of chemical dissection, Zwilling confirmed the finding first reported by Saunders[231] that the apical ectodermal ridge (AER) is indispensable for normal outgrowth of limb rudiments. (For review see Zwilling.[313])

Specificity of AER action

When the apical ectoderm from the leg of a chick embryo was transplanted to the limb bud of the wing after the latter had been deprived of its own apical ectoderm, the distal structures that formed from such a combination were characteristic of the wing.[234] Conversely, when the hind limb mesoderm was denuded of its own ectodermal cap and subsequently covered with apical ectoderm that had been removed from the wing bud, the structures formed from such a combination were of the leg type. These results suggested that the major determining component resides in the mesoderm, and that the ectodermal ridge acts as an inductor alerting the mesoderm to initiate processes of differentiation according to instructions that must be in the mesodermal cells. (For a review see Saunders *et al.*[234])

Limb formation in amphibians

Most of the older experiments on amphibians had also pointed to the mesoderm as the determining factor of limb architecture.

Harrison[138] had performed a famous experiment in which limb mesoderm was removed from the limb bud, leaving only the ectoderm *in situ*. In the absence of mesoderm, the remaining ectoderm was unable to carry out limb formation. The independence of the amphibian limb disc was furthermore attested to by classical experiments in which it was shown that when limb mesoderm was transplanted to a strange environment and placed underneath nonlimb ectoderm, a limb would form. On the other hand mesoderm derived from nonlimb sources did not form limb when placed in contact with limb ectoderm.

The results of these classical experiments suggesting that the ectoderm contributes no specific qualities to the differentiating limb, seem to contradict the modern experimental data of Saunders,[231] Saunders and Gasseling[233–234] and of Zwilling.[310, 313] They may be reconciled, however, by assuming that in these earlier experiments the limb mesenchyme that was tested had already been acted on by the ectodermal ridge.

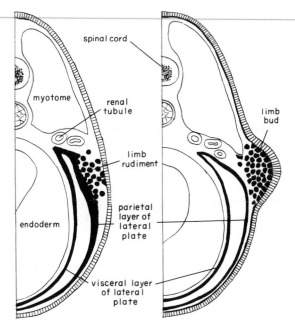

Fig. 4.16   Diagram of the origin of limb mesoderm in an amphibian embryo. (From Balinsky[10])

Timing of the inductive capacity of the apical ridge

Evidence for the importance of timing in assessing the role of the apical cap as inductor for limb outgrowth came from experiments by Zwilling[311, 313] in which the apical ectodermal ridge (AER) was removed by incubating the limb bud in a solution of the chelating compound ethylene diamine tetraacetate (EDTA) and then grafting the wing mesoblast to the flank. No distal outgrowths from such denuded limb mesoblasts were ever obtained. Distal outgrowths from EDTA-treated limb mesoblasts could be achieved, however, after they were again covered with a new apical ectodermal ridge. There is considerable evidence that the initial appearance of the ectodermal ridge itself is a consequence of mesodermal action. If *prior* to the appearance of the limb bud, the prospective wing-forming zone of the mesoderm is grafted to the belly side of the embryo flank, the ectoderm which will be made to heal over such a graft will form an apical ridge, and the mesoderm at its new site will grow out to produce a distal wing structure. Timing is apparently most important for this interaction to occur. Wing mesoderm quickly loses the power to induce an apical ridge in flank ectoderm shortly after the wing bud appears and flank ectoderm loses its capacity to respond to wing mesoderm at about the same time.

Mutants affecting limb formations as models for the analysis of gene effects on inductive action

The model of induction provided by the interplay between the apical ectoderm and the wing mesoderm is particularly suitable for analysis of the genetic control exerted on the inductive process because there exists in chicks a variety of mutants leading either to winglessness, polydactyly, syndactyly and other abnormalities of limb structures. Zwilling[310, 312] first turned his attention to the analysis of the mutation 'winglessness' in chicks. Winglessness is a simple recessive Mendelian mutant leading to complete absence of the wings in chicks. Zwilling noted that the limb buds of embryos homozygous for the mutation 'wingless' lacked an apical ectodermal ridge. In embryos homozygous for this mutation the ectoderm covering the wing mesoblast remained a flat sheet and never formed the characteristic conspicuous cap.[312]

Fig. 4.17   Right leg of 9½-day polydactylous embryo. The duplication involves only digit I (arrow) ×8. (After Goetinck[96])

Reciprocal recombination of wild type and mutant limb bud components

In a series of subsequent experiments by Zwilling[312] the mesoderm of limb buds obtained from normal wild type chick embryos was combined with ectoderm derived from wingless chick mutants. No limb was formed. Next, when mesoderm from the wingless mutant was grafted underneath the apical ectoderm of a normal wild type chick, the apical ridge covering the mutant mesoderm planted underneath it flattened out and regressed within three days. Development of distal limb structure was inhibited.

These experiments provided the first clue that perhaps the inductive relationship between the apical ectoderm and the limb mesenchyme is a two-way street. The apical ectodermal ridge induces the mesoderm to form distal limb structures. The apical ectodermal ridge in turn depends

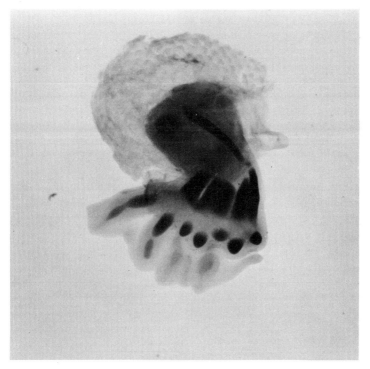

Fig. 4.18   Left leg of 9½-day talpid embryo. Seven digits are evident in this syndactylous foot. × 10. (After Goetinck[96])

for its continued maintenance on a factor emanating from the mesoblast, the so-called apical ectodermal maintenance factor (AEMF).

Zwilling[312–313] postulated the existence of an ectodermal maintenance factor (AEMF) which is responsible for this feedback, and which can be destroyed as a result of mutation. This hypothesis was further confirmed by results obtained from reciprocal recombinations of normal limb bud components and components obtained from the dominant polydactylous mutant.

In the *polydactylous* phenotype one or more extra toes are formed at the extremities. Limbs developing from combinations of tissues derived from the dominant polydactylous mutant with components from the normal wild type gave rise to the polydactylous phenotype only when the mesoblast was of polydactylous origin. The gene 'dominant polydactyly' in both homozygous and heterozygous condition leads to extreme hyperplasia of the apical ridge in the preaxial part of the limb, which is precisely the site where supernumerary digit formation takes place.[314]

A situation analogous to that described for dominant polydactyly has been reported to occur in the *talpid* mutant.

Embryos homozygous for the gene talpid are characterized by the presence of seven to eight syndactylous digits per foot. Studies of limb morphogenesis revealed that the limb buds of talpid are covered with an apical ectodermal ridge far more extensive in volume than that of normal wild type chickens. In recombination experiments between limb bud components obtained from this mutant and from normal wild type embryos the talpid phenotype is always obtained when the mesoblast is derived from the mutant genotype. The genotype of the limb epidermis does not contribute to the phenotype. Goetinck[96] proposed that perhaps the distribution of the AEMF was affected in embryos homozygous for this mutation.

The third case of exceptional limb outgrowth is found in the *eudiplopodia* mutant. This mutant is characterized by duplication of distal leg parts. In this mutant, a supernumerary ridge (AER) develops on the dorsal epidermis of the limb bud, a site where normally an ectodermal ridge never develops. Goetinck[96] suggested that in this mutant a portion of the dorsal epidermis becomes sensitive to the mesoblast stimulus and responds by forming an apical ridge. As a consequence, extra digits develop. The description of these mutations, and analyses of the results of recombination experiments certainly show that the inductive interaction between the ectoderm and the mesoderm can be affected by gene mutations.

We know nothing of the action of these mutant genes in molecular and biochemical terms but the chances seem good that these mutants are prepared to tell us more if we learn to ask the right questions.

SOME FINAL SPECULATIONS ABOUT THE MECHANISMS OF INDUCTION

In spite of all the information accumulated so far, the answer as to how the inductor acts on the competent cell continues to evade us. Two alternative conceptual models have, however, been proposed, and these are graphically presented by Wolff[299] in Fig. 4.19.

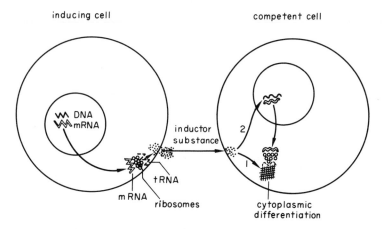

Fig. 4.19   Diagram presenting two theories of the action of inductors on a competent cell. (1) Short circuit. (2) Long circuit. (From Wolff[299])

According to these models then the inductor releases a preexisting cytoplasmic substance from some inhibitor. One of the models implies that the inductor acts by activating or derepressing a gene in the competent cell. The other implies that the inductor could react with a cytoplasmic factor and from such interaction a substance capable of activating or derepressing genes could be released.

Which theory is to be preferred? Wolff[299] defends the second alternative on theoretical grounds. He considers it *a priori* unlikely that so many different potential reactions are virtually present in the cytoplasm.

In addition there are a few inconclusive experiments that also tend to favour the second alternative. Brachet and Denis[43] and Denis[77] have shown that pieces of ectoderm treated with actinomycin D for a few minutes prior to association with normal organizer (blastopore lip) fail to form neural tube. If, as is generally assumed, actinomycin D interferes with synthesis of mRNA their results would favour the theory that inductive action influences genetic transcription and that tissues in which transcription is blocked, lose their competence to respond to inductors. It will be noted that in Wolff's[299] diagram the inductors could react with the cytoplasm of the competent cell and still be capable of activating or derepressing a gene.

Whatever the interpretation that is finally adopted, it seems likely that activator substances may be far more universal in embryogenesis than at one time thought, and may not be restricted to the classic cases of early induction. Diffusible substances may be transferred from one group of cells to another in early embryos, where short distances allow for diffusion to be effective. Subsequently *hormones*, i.e., substances that are by definition carried to the target cell via the blood stream, may continue to time the onset of reactions in the developmental sequence.[134] (For a review of hormone action in development see Hamburgh.[133])

5

Gene Action and the Fate of Informational Molecules during Development

DEVELOPMENTAL GENETICS

Implicating the nucleus in the direction of morphogenesis

A generation of biology students raised and nurtured on molecular biology may find the question 'Are genes operating in development?' about as relevant as the problem of whether a circle is round. It is a popular cliché, frequently repeated at perorations of scientific conventions, that the solutions one obtains to any problem depend on the questions one asks. Strange as it may seem, the question of whether and how genes act in development had not been asked by many biologists until relatively recently.

The classic period of embryology right through the 1920s and 1930s saw embryologists largely preoccupied with the organizer and the problem of induction. In fact it seems to this writer that embryologists never quite got over the organizer and to this day devote perhaps an excessive amount of attention to it. During the same period geneticists were essentially interested in the mechanics by which genetic factors are transmitted from parents to offspring.

To be a historian is even more difficult than to be a biologist and to attempt to assign priorities in the history of science invites accusations of inaccuracy. However, it is reasonably safe to assert that among the earliest pioneers who directed attention to the possibility that developmental processes were under nuclear control were Boveri and Hämmerling.

Boveri's merogonic hybrids

In a classic experiment, T. Boveri removed the nuclei from the eggs of sea urchins of one species and artificially fertilized them by sperm of another species with markedly different characteristics in pigmentation, shape of the pluteus larva and arrangement of the water vascular system. Such **merogonic** hybrids always resembled the species that furnished the nucleus, i.e. the sperm, suggesting that nucleus directs development and structural differentiation, and that the surrounding cytoplasm exerts no fundamental guidance on morphogenesis. In the phraseology of modern biology, Boveri's experiment might be listed as the first demonstration that the nucleus gives out the information according to which the 'construction' of the embryo proceeds. It was Boveri's luck that he concentrated on those features of the embryo, the ontogeny of which is not directed by long-lived messengers, previously prepared by the oocyte nucleus and stored in the egg cytoplasm (see Chapter 2, p. 33). Had he done otherwise, he might have been led quite astray. The work of Hadorn, for instance, has indicated that for certain traits in two species of *Triton*, the male nucleus of the fertilized egg is without effect on some tissues of some merogonic hybrids.

Hämmerling's experiments

The other early example that provided evidence that control of developmental processes is exercised through influences emanating from the nucleus, and, therefore, from the genes came from observations reported by Hämmerling on *Acetabularia*.[136] *Acetabularia* is a large unicellular alga consisting of three parts, a **rhizome** or hold-fast, a stalk, and a **cap** or umbrella. The nucleus of this single large cell is located in or near the rhizome. Hämmerling showed that when the stalk and the cap are cut off, the rhizome could regenerate the missing parts of the plant, provided the nucleus was left intact. The specificity of the directive influence exerted by the nucleus was demonstrated by the following experiments. Two species of *Acetabularia*, *A. mediterranea* and *A. crenulata*, are distinguished by the shape of their cap. In grafting experiments, in which the stem of *A. crenulata* was transplanted to the rhizome of *A. mediterranea*, a new cap was eventually formed, but the shape of the cap formed was intermediate between the phenotype of *crenulata* and *mediterranea* in the first regenerate. When the stem of this hybrid organism was again cut off and grafted a second time to the rhizome of *A. mediterranea* the new cap that formed was entirely of the *mediterranea* phenotype. Although the cytoplasmic material was provided by the stem of *A. crenulata*, the resulting shape of the cap was determined by the nucleus of *A. mediterranea*. Reciprocal experiments gave the same results (see Fig. 5.1). These findings were interpreted by Hämmerling to indicate that morphogenesis is controlled by the genome of the nucleus. (For review see also Brachet, Denis and de Vitry.[43])

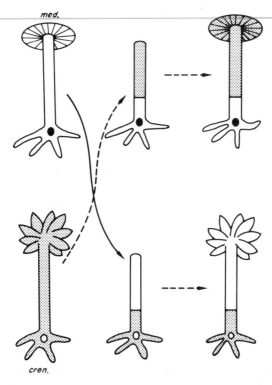

Fig. 5.1 Acetabularia, a large unicellular alga with a single basal nucleus. The morphology of the cap differs slightly in two related species. If the stem is cut, regeneration of the cap will occur only if the nucleus is retained. If a cap and stem is cut off from *mediterranea* basal portion and in its place a *crenulata* stem is grafted, the cap that is regenerated will be characteristic of the *mediterranea* species. The *mediterranea* nucleus will dictate to the *crenulata* cytoplasm the morphogenesis of the cap. (From Kerr[166])

The developmental analysis of gene mutations affecting morphogenesis

Identifying deviations from normal development

Whenever we ask the question how genes produce an observable effect, we are raising a problem of development. Developmental genetics is concerned with the analysis of the mechanisms by which genes produce their observable effects. Richard Goldschmidt may, perhaps, be called the father of physiological and developmental genetics, for at a time when most geneticists were busily constructing linkage maps, he posed, for the first time, the question. 'How do genes act in development?' Goldschmidt himself proposed that genes influence development by controlling the rates of physiological processes, a notion that has since been abandoned. He stimulated a small group of geneticists, however, to seek the answer to the problem of the mechanism of gene action, by investigating the developmental sequence by which certain easily identifiable single Mendelian gene mutations lead to major structural abnormalities. Reference is made to the now classic work of Bonnevie,[26] Dunn,[83] Gluecksohn-Schoenheimer,[93–95] Hadorn,[128–129] Landauer,[170] Poulson,[217] Rudnick,[225] Russell,[228] Waddington,[273] and Zwilling.[308–309] This approach is still continued by a number of younger embryologists and developmental geneticists employing the more modern and sophisticated techniques that have become available for the analysis of gene mutations: Auerbach,[2] Bennett,[23] Carter,[58] Dagg,[69] Hamburgh,[131–132] Smith,[239–240] and Wittman and Hamburgh.[298] The mouse and, to a lesser extent, the chick were the favourite animals on which analyses of the developmental sequences by which mutant genes interfere with normal development were carried out. Both the mouse and the chick represented a compro-

Table 5.1. Developmental effects of selected mutants in the mouse

Mutation	Phenotype	Developmental System Affected
Brachy (T)	Short tail or absence of tail	Notochord Neural tube Tail structures
Kinky (K)	Kinky tail or curved tail	Notochord Somites Neural ectoderm and mesoderm
Fused (Fu)	Fused tail ankylosis between vertebrae	Neural tube Notochord
Heminelia tibiae or luxate (lx)	Complete reduction of tibia and bent fibula	Preaxial mesenchyme of limb bud
Danforth's short tail (Sd)	Short tailed in heterozygous condition Absence & reduction of kidney in homozygotes	Tail notochord, tail somites Tail vascular systems Ureteric bud Nephrogenic mesenchyme
Reeler (rl)	Growth retardation Ataxia, Tremors	Fetal granular cells of cerebellum Hippocampus Cerebral cortex
Quaking (qk)	Tremors and Convulsions	Cerebral myelin

mise between the demands of the geneticists and embryologists because these two organisms have been studied by both geneticists and embryologists, and sufficient amount of information was available for the two disciplines to join hands.

Some of the developmental effects of gene mutations on developmental processes are listed in Table 5.1, opposite.

The descriptive recording of the deviation from normal development in selected mutants of the mouse and chick have indeed revealed that practically every embryonic process, including induction, organizer activity, mitotic rate, proliferation, growth, cell migration, resorption, selective cell death, tissue aggregation and disaggregation is under genetic control. (For reviews, see Glucksohn-Waelsch[95] and Grüneberg.[119–120]) The information gained from this approach added con-

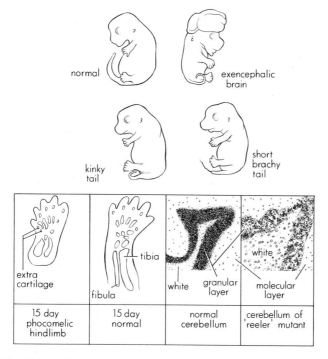

Fig. 5.2 Schematic representation of some genetically determined malformations in mice. Top row shows the mutation exencephaly described by Bonnevie.[26] Below the brachy mutation analysed by Bennett,[23] Grüneberg[120] and Wittman and Hamburgh.[298] Bottom row shows developmental stages in the mutation phocomelia investigated by Sisken and Glucksohn-Waelsch[238] and the reeler mutation analysed by Hamburgh.[131] [132]

vincing proof, if any was needed, that genes play a decisive role in the chain of events occurring during development and that mutant genes interfere with normal processes of differentiation.

The major generalizations from developmental studies of gene mutations

In addition, a number of generalizations have emerged from studies of developmental genetics.

DISTINCTION BETWEEN AUTOPHENES AND ALLOPHENES

One question that always arises in a study of developmental effects of genes is whether the damage exerted by a mutant gene is due to an intrinsic effect on the affected cells (*autophenes*) or whether the gene effect is extrinsic (*allophenes*) to the cells whose normal development is interfered with. Analysis of the developmental pathways of mutant genes have yielded examples of both.

In the creeper fowl, the dominant factor *cp* in heterozygous dose causes a typical chondrodystrophy of the long bones. In *cp/cp* homozygotes, embryonic development stops at 96 hours and death results in 90 per cent of the cases. The remaining 10 per cent escape the lethal effect of the gene at this stage, but succumb at a later time.

Transplants of leg rudiments of *cp/cp* homozygotes (carried out by Hamburger and Rudnick) into normal hosts manifest the same degree of phocomelia as do the limbs in the inviable homozygous embryos. The *cp* gene may, therefore, be classified as a true autophene. On the other hand, the *cp* gene in addition to the disturbance it causes on limb differentiation causes abnormalities of eye development, such as microphthalmia and anophthalmia and folding of the retina.

An eye rudiment of *cp/cp* homozygotes transplanted to a normal host develops normally in the healthy environment. Conversely, an eye rudiment obtained from a normal embryo and transplanted to a *cp/cp* host developed the abnormal pattern typical of *cp* homozygotes. With respect to eye development, the same *cp* gene acts, therefore, as an allophene.

PLEIOTROPISM

The example cited above points up another major generalization obtained from these studies, namely the observation that a given genetic factor may exert multiple or *pleiotropic* effects. *A priori* there is no good reason why a given gene should not have multiple effects on entirely unrelated structures. In an organism like *Drosophila*, where a conservative estimate postulates 10 000 genetic loci, one would have to assume that a given gene codes for more than one structure or trait, or that the same code exerts varying effects in different cell populations. On the other hand, many instances of apparent pleiotropism may merely reflect differential sensitivity among a variety of cells and tissues.

Genes that code for proteins or enzymes that are shared by all cells, such as respiratory enzymes, would appear to exert pleiotropic effects in contrast to those loci that code for tissue specific proteins.

Before one can assign true or 'genuine' pleiotropic effects to a gene, one must first rule out that the multiple effects on divergent cells and structures are not related in a hierarchy of causes, a situation for which Grüneberg has coined the term *spurious pleiotropy*. Grüneberg has revealed multiple effects of a number of genes on divergent tissues which appear to be unrelated but can be fitted into a pedigree of causes.

One of the most celebrated examples of this type of spurious pleiotropy is the mutation *myelencephalic blebs* in the mouse. The primary deviation from normal developmental sequence in this mutant is the leakage of cerebro-spinal fluid from the foramen of Magendi. The movement of these blebs along the body axis causes severe disturbances at all those sites of the body where the fluid-filled blisters accumulate. Accumulation of blebs near the eye causes eye defects, while accumulation near the limb buds leads to foot anomalies.

CRITICAL PERIODS OF GENE ACTION

Analysis of development of mutants does not reveal the true timing of onset of gene activation, but merely the time when the gene product that has been elaborated begins to take effect on differentiating cells. It is obviously more difficult to make statements about timing when observable effects occur late in development, than when they occur early in the developmental sequence.

The best example of how long the delay between elaboration of gene product and observable effect can be, are the so-called *maternal effects* where genes present in the cell of the female or the premeiotic egg cell cause developmental effects in the offspring, even though the genes may not have been included in the zygotic nuclei.

On the other hand, there are methods available which can demonstrate onset of gene action very early in ontogeny, and a number of these methods permit one to establish the earliest embryonic events influenced by gene action.

Poulson[217] has shown that several chromosomal deficiencies lead in some cases to very early and in other cases to very late arrest of development in *Drosophila*. The white locus in *Drosophila*, which can be recognized by the removal of 10 bands in the giant chromosome of the salivary gland, is an example of a gene causing late arrest, for it permits completion of differentiation of ectoderm prior to its disturbing action. The scute deficiency (11 bands removed in the giant chromosome) permits development into fully differentiated larvae which subsequently die at hatching. In the notch deficiency, on the other hand, development comes to a stand-still already at 6 hours of embryonic life. The study of the development of *merogonic hybrids* permits some more precise estimate of the timing of gene action in development. In such hybrids, the enucleated cytoplasm of one species is combined with the sperm nucleus of another. Thus, the egg cytoplasm of *Triton palmatus* can be made to develop with the nucleus of *Triton taeniatus* to late larval stage; but when *T. palmatus* cytoplasm is combined with the nucleus of *T. cristatus*, development stops shortly after neuralization. Possibly, in two 'incompatible' species, the genetic loci required by a given cytoplasm are not present in a foreign nucleus. This may not mean that the genetic locus begins to act precisely just at the time when the visible effect becomes noticeable, but that an RNA may be produced which is not compatible or able to act in the foreign cytoplasm. When the lack of proper RNA codes becomes critical, normal development is no longer possible.

Zeller, working on the early lethal merogonic combination of *T. palmatus* cytoplasm with *T. cristatus* nucleus, found this system able to synthesize normal amounts of RNA. Possibly, the RNA produced by the *cristatus* nucleus and released into the *palmatus* cytoplasm might lead to an accumulation of RNA molecules that are different in one or more respects from that which a *T. palmatus* nucleus would have provided. In fact, in this merogonic combination, RNA accumulates to an abnormally high level with the onset of the lethal crisis.

The phenocopy concept

One of the most interesting approaches to the problem of gene action involved the production of so-called 'phenocopies' and the subsequent attempts to learn more about mechanisms of gene action by studying the effects of experimental agents that mimic the effects of mutant genes.

At the time when many embryologists (and teratologists) demonstrated the variety of malformations that could be produced in chick and mammalian embryos by experimental procedures, some geneticists were struck by the similarity of these malformations to the abnormalities caused by single Mendelian gene mutations.

The term *phenocopy* which was invented in 1935 by Goldschmidt was based on the observation that exposure of *Drosophila* larvae to high temperature brought on modifications from normal development, which resembled a large array of mutant phenotypes. To Goldschmidt, the probability that this parallelism is due to chance, appeared negligible. Landauer subsequently expanded the criterion for true phenocopies, requiring that the critical stage in development during which a particular modification is experimentally produced should coincide with the developmental stage of the homologous mutant. Step by step alterations which lead to a genetic variation should be paralleled by the experimentally produced copy of that variant. Landauer suggested that for a true phenocopy an additional criterion would be the demonstration that both the penetrance and the expressivity of the gene to be copied, and the response to the environmental agent, leading to similar or identical phenotypic end effects, are subject to the same genetic modifiers. In a series of papers, Landauer[171–173] reported a number of teratogenic consequences resulting from the presence of insulin, 6-aminonicotinamide, sulphanilamide, and boric acid on developing chick embryos. The abnormalities induced by these agents resemble a number of known and well analysed gene mutations such as 'rumplessness', 'micromelia', 'parrot beak', and 'crooked neck'. In each instance, Landauer could show that the similarity between the experimental modification and the mutant was not a superficial resemblance, but included identical developmental pathways and response in the same direction to the same genetic modifiers. His experimentally induced malformations, therefore, fulfilled all the requirements that he imposed on the phenocopy concept. (See also Dagg,[69] Hamburgh,[130] Hamburgh and Callahan.[135]) The phenocopy concept, so defined, held high hopes for the developmental geneticist.

'Since it is obviously very difficult to analyze individual links in the chain of processes leading from gene to phene . . . it would be of great advantage to be able to produce artificially an abnormality, which not only in its appearance but also in the details of its mode of origin is identical with a gene-produced abnormality. If such a true phenocopy could be manufactured it would perhaps be possible to get closer to the analysis of primary gene effects and the fundamental problem of mechanism of gene action.' (Gluecksohn–Waelsch[95]).

None of the deleterious agents studied so far either in chicks or in the mouse proved to have primary effects exactly identical with a gene-produced abnormality, in spite of the resemblance of the routes by which the end results came about. This was most clearly demonstrated in Landauer's example of insulin-induced rumplessness in chick embryos,[171] which resembles the recessive gene for rumplessness both in developmental pathway as well as the direction by which genetic modifiers influence the penetrance and expressivity of this gene and the severity of the response to insulin. Insulin-induced rumplessness, however, could be reversed by simultaneous administration of nicotinamide. This substance proved totally ineffective in alleviating the abnormality when fed to chick embryos homozygous for the 'rp' gene 'rumplessness'. The hope that among a multiple step sequence of reactions between initial gene effect and the final end result, the identical step in a chain of events would be interfered with by an environmental agent and a gene, even though the former may mimic the latter, is based on a highly improbable assumption.

RNA METABOLISM DURING DEVELOPMENT

Introduction and historical review

Stimulating as the classical approach of developmental genetics may have been, these studies did not really provide satisfactory answers to the problems of how genes act in development.

In the light of the new theories of modern molecular biology, according to which RNA molecules serve both as codes and as the machinery for synthesis of protein molecules, information about the changes in RNA metabolism taking place during embryogenesis gained new significance for theories of development. If visible cell differentiation is preceded by biochemical changes that lead to the accumulation of those tissue-specific proteins which cells must elaborate in order to carry out their specific assignments, then the key to differentiation is to be found by exploring the mechanism by which cells are switched into one or the other synthetic path. Since the programming equipment of a cell resides in the DNA of the nucleus and the various categories of RNA elaborated from nuclear DNA, viz. the soluble or transfer RNA, the ribosomal RNA and the messenger RNA, the analysis of RNA metabolism during growth and differentiation may provide the key that can perhaps unlock the riddle of differentiation. If the doctrines of molecular biology have any relevance for the study of development at all, then the major developmental events, i.e. oogenesis, cleavage, gastrulation, induction, cell differentiation, should be paralleled or preceded by changes in bursts of activity in DNA and RNA metabolism. Historically, the first approach to correlate RNA metabolism with developmental events was provided by studies of Brachet.[33] Essentially, these studies concentrated on recording either geographic changes in distribution or sequential changes in RNA synthesis at different stages of development.

In his earliest attempts, Brachet[33] identified RNA by a nonspecific staining procedure, using the basic dye toluidine blue which revealed the presence of basophilic granules, presumably RNA, in cells of different regions of amphibian embryos (see Chapter 3). On the basis of intensity of staining, it was proclaimed by Brachet that the dorsal lip cells, and later the invaginated chordamesoderm, were highest in RNA granules. Far more exciting than the mere recording of differences in regional distribution of RNA in the amphibian gastrula was the observation that during invagination there was a loss of such granules from dorsal lip cells with a corresponding increase of granules in the overlying ectoderm. All of this suggested the very attractive hypothesis that during induction RNA may actually pass from inductor to competent tissue. None of these early studies could distinguish, however, between the different categories of RNA which are synthesized during different developmental periods.

The functional categories of RNA that have since been identified are separated by methods which take advantage of their physical and biochemical differences. *Pulse labelling* is the technique employed to distinguish between different classes of RNA. A radioactive precursor of RNA (^{3}H uridine) is injected into the test animal, this being followed after a few minutes by injection of unlabelled precursor (*chaser*) to dilute out quickly the radioactive material.

Presumably brief exposure to the labelled precursor is followed by incorporation of the labelled nucleotide into RNA. After some time, in the absence of labelled precursor (the so-called *chase period*), the stable, radioactive form of RNA can be identified. This RNA is then extracted, purified and subjected to sucrose gradient zonal centrifugation. This technique separates molecules by virtue of their size. The RNA is then layered over a linear gradient of sucrose solution and centrifuged at high speed, until molecules of different size sediment out as distinct bands. In animals the two *ribosomal* classes of RNA (rRNA) have sedimentation constants of 28S and 18S, while *soluble or transfer* RNA (tRNA), the lightest, has a sedimentation constant of 4S. A third class of RNA, the so-called *messenger* RNA (mRNA), is considerably less stable and is somewhat between these two in weight.

Synthesis and utilization of different classes of RNA during embryogenesis

Synthesis and utilization of mRNA in amphibian embryos

Studies employing these methods (i.e. using pulse labelling with radioactive nucleotide and subsequent sucrose gradient zonal centrifugation) have since revealed that the three classes of RNA are synthesized at different periods of embryogenesis. In amphibians the sequence of events can be summarized as follows. During oogenesis the synthesis of ribosomal RNA is extremely intense, while synthesis of tRNA is practically absent. Oogenesis probably constitutes also a period of active messenger synthesis. This conclusion is based on autoradiographic studies by Gall[91] and Gall and Callan.[92] They showed that there was a high rate of chromosomal RNA synthesis in lampbrush chromosomes of oocytes, and that this can be inhibited by actinomycin D, a substance which in appropriate concentration specifically stops the synthesis of mRNA. The mature oocyte, on the other hand, synthesizes little, if any, RNA. Synthesis of mRNA and tRNA is resumed again during cleavage (see Chapter 2, p. 34), but ribosomal RNA synthesis does not begin in earnest until the onset of gastrulation. During this stage, mRNA synthesis rises to a maximum peak. (For a review see Brown,[52-53] Denis[78] and Davidson.[75])

In agreement with Brachet,[39] it is not unreasonable to interpret the intense incorporation of uridine into nuclear RNA during gastrulation, and the sensitivity of the amphibian gastrula and neurula to actinomycin D, as signalling that gastrulation is a period during which mRNA and rRNA is synthesized in massive proportions and probably also utilized for transcription.

Synthesis and utilization of ribosomal RNA in amphibian embryos

A similar delay between synthesis and use occurs with respect to ribosomal RNA. In the cleaving amphibian egg, synthesis of ribosomal RNA does not occur until the time of gastrulation. The early events in development, such as cleavage and blastula formation, take place in the absence of ribosomal synthesis. Delay in ribosome formation was first demonstrated in amphibian embryos by Brown and Caston[54] who noted that in order to raise *Rana pipiens* embryos to feeding stage, it is necessary to supplement the medium with Mg^{++}. Mg^{++} ions, besides being a major requirement for many metabolic systems, are also required for synthesis of ribosomal RNA. The ions must be added when the embryo has reached the tailbud stage in order to insure development into a free-swimming larva. This late onset of Mg^{++} requirement by the amphibian embryo suggested to these authors[54] that the developing embryo does not need new ribosomes to carry it through the early period of differentiation. Since, however, protein synthesis must commence soon after fertilization, it is reasonable to assume that old ribosomes, manufactured presumably in the oocyte, might be used up during the pre-gastrulation stages of development.

The discovery of Elsdale, Fischberg, and Smith[85] of a mutant in *Xenopus* which possessed no nucleoli and was, therefore, incapable of synthesizing ribosomal RNA, provided an ideal system to analyse the onset of synthesis and the utilization of ribosomal RNA, in amphibian embryos. The homozygous mutant (O nu) develops normally at first but its development is arrested at the swimming stage with death following soon after.

In experiments performed by Brown and Gurdon,[55] the ability of (O nu) eggs to form rRNA was tested by growing control and mutant embryos (neurula and tailbud stages) in medium containing radioactive nucleic acid precursor ($^{14}CO_2$). The RNA formed was then purified and subjected to sucrose gradient centrifugation for identification. In the (O nu) embryos, only the transfer RNA was radioactive in contrast to

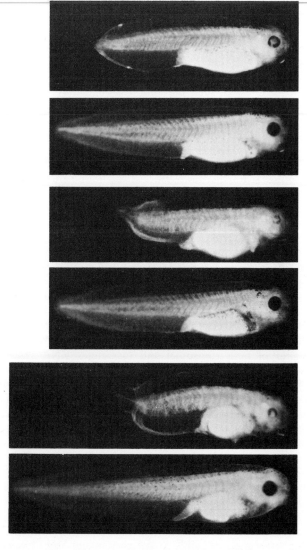

Fig. 5.3 Control and magnesium-deficent embryos of *Xenopus laevis*. Sibling embryos developed in a defined salt solution (Brown *et al.*[54]) with and without 0·1 mM MgCl$_2$. Alternate photographs from left to right show the developmental progress of the magnesium-starved and control embryos at daily intervals. (From Brown[52])

the control embryos in which all categories of RNA exhibited radioactivity. This suggests that control neurula and tailbud stages are actively synthesizing transfer, messenger and ribosomal RNA while the O mutant synthesizes only soluble RNA. The bulk of the RNA formed by controls was rRNA, but no new ribosomal RNA was formed in the mutant (O nu) neurula and tailbud stages. Since, however, development proceeded normally in such mutants, although they failed to form any RNA besides transfer RNA, it would appear that the ribosomes that are normally manufactured during neurula or tailbud stages are not utilized immediately but are stored to satisfy requirements for later events in differentiation (post-swimming stage). Protein synthesis that is taking place during the neurula or tailbud stages in turn seems to rely on rRNA previously made. Furthermore, since it was found that the total amount of rRNA of the mutant was equal in concentration to that found in the unfertilized egg, it was concluded that all the ribosomal RNA used by the mutant during the early stage was derived from conserved material preformed in the oocyte.

Synthesis and utilization of messenger RNA in echinoderm embryos

The question may be asked how universal is the presence of long-lived messengers in developing systems. To answer this question, a number of embryologists examined this problem in echinoderm eggs.

The pattern of RNA synthesis in echinoderms exhibits similarities to that in amphibians. During oogenesis, the oocytes of most echinoderms are the sites of both ribosomal and messenger RNA synthesis. (For review see Davidson.[75])

Ficq (Grant[99]) studied the effect of actinomycin D on incorporation of labelled cytosine by means of autoradiography in oocytes of *Paracentrotus* and could demonstrate that incorporation of labelled precursor was inhibited in oocytes exposed to this antibiotic. Protein synthesis continued undisturbed in such oocytes, suggesting that during oogenesis protein synthesis is not dependent upon simultaneous nuclear synthesis of mRNA.

In Chapter 2, p. 32, a series of experiments by Gross[114-115] and Gross and Cousineau[116-117] were described, that were designed to show how protein synthesis is regulated at fertilization.

In these studies, Gross[114-115] and Gross and Cousineau[116-117] have shown that cleavage is possible in the presence of high doses of actinomycin D in fertilized echinoderm eggs. As a matter of fact, dosages of actinomycin D, sufficient to inhibit 95 per cent of uracil incorporation, will not suppress cleavage or protein synthesis in cleaving eggs. It seemed, therefore, of interest to determine at what stage of development an embryo actually begins to activate the program for the synthesis of new and specific proteins and what effect, if any, interference with RNA synthesis exerts on processes of differentiation. The discovery that actinomycin D inhibits specifically the synthesis of DNA-dependent mRNA was received with tremendous interest by developmental biologists, and experiments testing the effect of actinomycin D on developing systems at every conceivable stage became almost routine. As pointed out in Chapter 2, p. 31, the various studies by Gross and Cousineau, Wilt, Hultin, and Nemer have since revealed that in sea urchin eggs ribosomes and transfer RNA are present in the unfertilized ovum. Messenger RNA also seems to be present in the unfertilized egg. It has been hypothesized that the messenger RNA of unfertilized eggs cannot serve as a template until, in the act of fertilization, some repressor protein is removed from the mRNA molecules. Although the fertilized echinoderm egg is making mRNA as well as protein immediately after sperm activation, Gross and Cousineau[116–117] showed that the messenger molecules that are used for protein assembly during cleavage are probably not identical with the burst of messengers that are synthesized immediately after fertilization. The almost instant inhibition of RNA synthesis that follows when freshly fertilized eggs are treated with actinomycin D does not interfere with the rise in protein synthesis that is taking place in the echinoderm eggs within minutes after fertilization. Actinomycin D applied at fertilization does, however, interfere with the second rise in protein synthesis that normally begins at gastrulation and reaches its peak at neurula stages.

These results have been interpreted to indicate that an important class of new mRNA is elaborated after fertilization, but is held in store and put to use after gastrulation. The presence of preformed stored mRNA molecules in unfertilized eggs is further supported by some additional observations, listed below:

1. Nonnucleated fragments of sea urchin eggs that are parthenogenetically activated are able to cleave and form blastulae. It is reasonable to suppose that since the nuclear sources of templates are absent, the required codes for the proteins of cell division (mitotic spindle proteins) must have been preformed and stored in the egg cytoplasm, or mitochondrial DNA must provide the templates for messenger molecules that code for mitotic proteins.

2. Application of actinomycin D to fully fertilized eggs stops new mRNA synthesis within minutes, but such eggs will continue to cleave and reach the blastula stage. The upsurge in protein synthesis started at fertilization is no less intense in actinomycin D treated eggs than in controls. Deviation from normal development in actinomycin D treated eggs is not initiated until gastrulation.

Based on these observations, Gross and Cousineau[116–117] proposed the hypothesis that the sea urchin egg contains preformed mRNA that is stored in the cytoplasm of the egg and not put to use until cleavage and during the blastula stage. Conversely, the utilization of mRNA that is actually synthesized in the post-fertilized egg, is probably delayed until gastrulation. (For a review, see also Grant.[99])

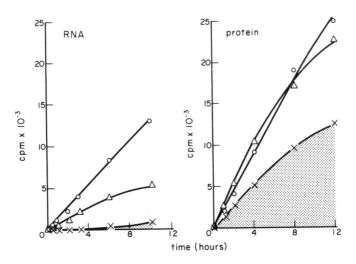

time (hours)

Fig. 5.4   Protein and RNA synthesis in *Arbacia* embryos exposed for different intervals to actinomycin D. After 4 hours of normal development, a suspension was divided in three: $\frac{1}{3}$ of the embryos were left to develop normally; $\frac{1}{3}$ were given actinomycin D (10 μg/ml) at 4 hours; $\frac{1}{3}$ were given actinomycin D at 24 hours. Penicillin and streptomycin were present from the beginning in all. At 24 hours, all suspensions received ^{14}C-uridine or ^{14}C-valine, and samples were taken over the next 12 hours. Counts in protein or RNA for 10^4 embryos are shown for: actinomycin D at 4 hours (×–––––×), actinomycin D at 24 hours (Δ–––––Δ), no actinomycin D (control) (O–––––O). (From Gross[114])

Similar results reinforcing the idea that synthesis of informational molecules precedes in time visible changes of differentiation were reported by Collier[61] on the spiral egg of *Ilyanassa*.

Some generalizations and conclusions

It is too early to present a general theory of differentiation on the molecular level. The observations showing that different classes of nucleic acid are synthesized initially at different developmental times, and that the stages at which these molecules operate in programming are divorced in time from the stage at which they are synthesized, have, however, suggested some hypotheses for further testing.

It seems to be generally accepted that during cleavage, while DNA is engaged in duplication, appreciable amounts of RNA cannot be synthesized simultaneously. If the observation is correct that the period preceding gastrulation is a relatively silent period with regard to the synthesis of rRNA and of mRNA, one would have to assume that the egg contains a programme of previously supplied information, that is sufficient to carry it through to the blastula stage.

The concept of a completely silent period does not, however, make too much sense, nor is it really in agreement with observed facts. No such silent period exists in mammals, where Mintz[195] showed continued synthesis of mRNA and rRNA starting with the morula stage.

In the sea urchin egg, DNA-like RNA synthesis, though low in concentration, has indeed been demonstrated as early as the four cell stage. In amphibians, onset of mRNA synthesis precedes gastrulation by several hours. As pointed out above, continuous synthesis of mRNA and rRNA has been demonstrated as early as the morula stage by Mintz[195] for the mouse embryo.

The universal pattern seems to be that preformed messengers, presumably made in the oocyte, operate to assure the synthesis of those proteins necessary for blastula development and that messenger molecules synthesized during cleavage stages probably operate to code proteins and morphogenetic substances that emerge during the gastrula stage.

6

Genetic Information and the Programming of Development

ARE NUCLEI EQUIPOTENTIAL?

Statement of problem

A priori there are only two ways out of the logical dilemma (see Epilogue) in which the student of development finds himself. The visible transformations that arise during development in an initially homogenous population of cells, such as the blastula, may merely reflect differences in kind and patterns of enzymes within the biochemical household of these cells. Since the synthesis of these enzyme proteins requires the elaboration of instructions by genes, and since all blastomeres contain the total complement of genes, it follows that either the genes themselves must undergo changes in the course of development as they get distributed into generations of descendant daughter cells, or very subtle changes are introduced whenever two or more neighbouring cells interact with one another, resulting in some sort of feedback which affects the activation or deactivation of genes, as a consequence of which gene products may become differentially distributed.

Nuclear transplantation experiments

The pioneer work of Briggs and King

It was the former hypothesis, namely the possibility that nuclei themselves may change as they come to reduplicate and inhabit subsequent generations of daughter cells, that was first investigated. The investigators who deserve the credit for devising the experiments to test the proposition of **nuclear change** during development were Briggs and King[46-50]. Other investigators, including, for example, Moore[201], Gurdon[121-125], and Fischberg *et al.*,[88-89] have confirmed and extended these studies. Briggs and King, acting in the tried manner of two detectives who attempt to identify the plan of operation of a suspected criminal, proceeded to subject their suspect, the nucleus of an embryonic cell, to a series of well-defined situations and then to test its responses. The suspects of Briggs and King were the nuclei of amphibian eggs and the nuclei of cells of the blastula, gastrula, early neurula and tailbud stages of frog embryos. Specifically, they wished to find out whether nuclei derived from blastulae, gastrulae, or cells of the tailbud stage of the amphibian embryo could be transplanted back into an enucleated egg and still support normal development. In short, the question they asked was: Are genes really *equipotential*? Their attack on this problem involved the following strategy.

Frogs were induced to ovulate several hours after they had received an injection of anterior pituitary glands. The unfertilized egg of *Rana* so obtained was activated by pricking with a glass needle. The egg nucleus, which lies just under the surface of the animal pole, was then removed, leaving an enucleated egg cell that could act as host. Nuclei were removed from either donor blastulae, gastrulae, or endoderm of tailbud stages by placing the donor tissue into saline medium that was lacking in Ca^+ and Mg^+ ions but contained 1×10^3 EDTA (ethylenedinitrotetra acetic acid disodium salt), an agent that promotes dissociation of cells. A prospective donor nucleus was selected and the donor cell was sucked into a micropipette of smaller diameter than that of the cell so that the cell membrane was broken. The nucleus surrounded by some cytoplasm was then injected into the animal pole region of the enucleated, activated recipient egg.

Using these procedures, Briggs and King showed that blastula nuclei of *Rana pipiens* are capable of promoting complete and normal development when introduced into **enucleated test eggs.** However, as development proceeds, there occurs a progressive restriction in the ability of somatic cell nuclei to promote normal development of enucleated test eggs. In high proportion and statistically significant numbers, endoderm nuclei obtained from late gastrulae and transplanted into enucleated frog eggs promoted only a few cleavage divisions.

Subsequent development was arrested at the late blastula or early gastrula stage. Among the few endoderm nuclear transplant embryos that had completed gastrulation, many were reduced in size or subsequently developed a variety of abnormalities affecting chiefly the central nervous system and sense organs, or neural crest derivatives. The severity of these abnormal changes increased if endoderm nuclei from midgut

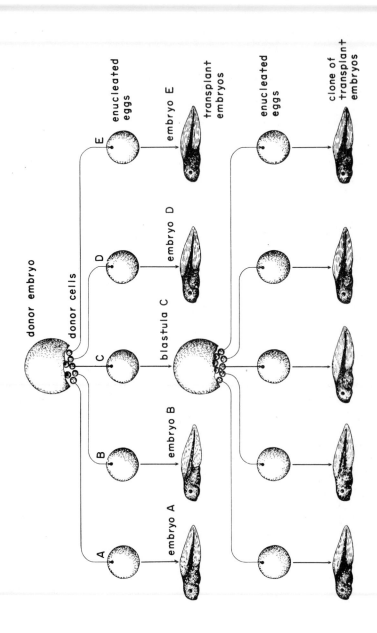

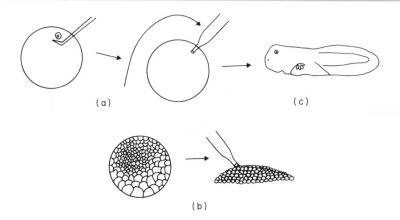

Fig. 6.1 Briggs and King's nuclear transplantation experiment. **(a)** A frog egg is enucleated and **(b)** the nucleus from a blastula cell of another embryo is injected into the enucleated egg. **(c)** The egg cytoplasm with a blastula nucleus develops into a normal tadpole. (From Kerr[166])

of a mid-neurula were used. Endodermal derivatives such as the alimentary tract, developing from eggs into which endoderm nuclei had been transplanted, were least affected.

From these results, the authors concluded that the capacity of endoderm nuclei to promote co-ordinated differentiation is progressively restricted. In the late gastrula, some endoderm nuclei are apparently still unchanged and totipotent, others are merely limited in their capacity to promote ectodermal differentiation, and still others are totally incapable of promoting either ectodermal or mesodermal differentiation. It is, of course, not surprising that, at any given stage of development, not all nuclei should be alike with respect to the degree of restriction that is imposed upon them.

Are nuclear changes irreversible?

The next question to which the two investigators addressed themselves

Fig. 6.2 Production of nuclear clones makes it possible to test various assumptions concerning the specialization of cells. Nuclei from donor cells of a partially developed frog embryo are injected into enucleated eggs, i.e. eggs from which the nuclei have been removed. These produce five normal transplant embryos. Nuclei from cells of blastula C are injected into another set of enucleated eggs. The resulting clone of transplant embryos is five genetically identical individuals whose chromosomes are all descended from chromosomes of one nucleus. (From Fischberg and Blackler[88])

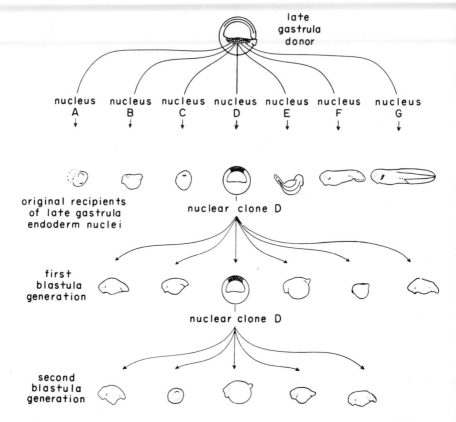

Fig. 6.3 Diagram illustrating serial transplantation of endoderm nuclei. Donor nuclei are actually taken from the presumptive anterior midgut region of the late gastrula. Transferred to enucleated eggs they promote the various types of development shown for the 'original recipients' in the diagram. One of the original recipients, sacrificed at the blastula stage, provides nuclei for a single clone which shows the more uniform development illustrated for the first and second blastula generations. In this and subsequent figures the illustrations of embryos are in the form of either camera lucida drawings or photographs. (From King and Briggs[168])

was whether the nuclear changes are stable and irreversible and whether they are specific. This problem was attacked by Briggs and King in the following manner: (See Fig. 6.2 and Fig. 6.3.)

Several endoderm nuclei of a late gastrula were transplanted singly into enucleated eggs (Fig. 6.3 A-G). A spectrum of malformations ranging from complete arrest (A) to normal development (G) was observed. Egg D was not permitted to continue development beyond blastula formation. Instead, several blastula nuclei, all of them derivatives from a single endoderm nucleus that had originally been selected for transplantation (to egg D), were transplanted singly into several eggs. Surprisingly enough, unlike ordinary blastula nuclei, they did not promote normal development, but the abnormalities in this new clone were all alike. When this procedure was repeated and one blastula from clone D was sacrificed to act as nuclear donor for another batch of eggs, the generation derived from these blastula nuclei was again almost identical with respect to their developmental limitations. Having ruled out changes in chromosome number or morphology in some of the sampled donors, these authors concluded that the nuclear restrictions emerging in the course of development and cell differentiation are relatively stable.

A very elegant experiment to prove that permanent changes can be imposed on embryonic nuclei was supplied by studies of Moore[201]. He transferred diploid *Rana pipiens* nuclei into enucleated eggs of *Rana sylvatica*. Development stopped at the blastula stage. When the nuclei from such blocked (*R. sylv.*) blastulae were reimplanted into *pipiens* eggs, development was again blocked. Apparently, exposure of the *pipiens* nucleus to the *sylvatica* egg cytoplasm changed the nucleus sufficiently so that even after it was returned to its own cytoplasmic home, i.e. the *pipiens* egg, it could no longer support normal development.

Recently, Di Berardino and King[80] reported parallel experiments to those of Briggs and King with transplanted nuclei derived from the presumptive neural plate area of the late gastrula and the definitive neural plate region of early and mid-neurula embryos of *Rana pipiens*. Their results confirmed the findings first reported by Briggs and King on the basis of transplantation of endoderm nuclei. They reported that in the course of neural differentiation there occurs a gradual progressive change which restricts the capacity of most neural nuclei to promote normal development of test eggs. The results also implied that the nuclear changes responsible for these developmental restrictions are stable and cannot be reversed by repeated exposure to the cytoplasm of the original test eggs.

Gurdon's challenge to the theory of nuclear change

The interpretation first advanced by Briggs and King[48–50] and more recently by Di Berardino and King[80], that in the course of differentiation there occurs a gradual, progressive, and stable change, which restricts the capacity of nuclei derived from differentiated cells to promote normal development of enucleated test eggs, has been challenged by Gurdon[121–125] on the basis of experiments with *Xenopus laevis*. Transplantation of nuclei from such highly differentiated cells as the ciliated intestinal epithelium into enucleated eggs of *Xenopus* promoted normal development into feeding tadpoles in 10 out of 726 cases. Gurdon argued that the developmental restrictions observed in *Rana* may, therefore, not reflect any 'stable change' of the nuclei proper prior to their transfer into test eggs, but that the developmental arrest observed in *Rana* embryos following nuclear transplantation may be caused by events initiated during the transfer process itself, or *older* nuclei may simply become less compatible to young egg cytoplasm. On the other hand, the limited number of normal embryos that Gurdon obtained from *Xenopus* eggs containing transplanted nuclei derived from highly differentiated cells, does not, in itself, invalidate the theory of nuclear differentiation.

The process of development may be compared to a race in which a multitude of players participate. At any given time, no two runners are at exactly the same point from the starting line. Some are far out, some further back, the majority probably cluster around a mid-point. Terms like blastula, gastrula, and neurula are obviously somewhat arbitrary assignments, when referred to the state of differentiation of cells or of nuclei. They tell the developmental biologist merely 'how far from the starting line' the majority of cells are at a given time, it always being understood that a population of embryonic cells, just like any population, contains a 'Johnny come lately', as well as a 'champion' that can break the speed record. The fact that some nuclei randomly obtained from a population of cells like those of the intestinal epithelium, can still promote normal development, may merely mean that a limited number of them have not yet advanced all the way in the race toward differentiation.

An alternative explanation for the situation found in *Xenopus* was proposed by Di Berardino and King[80], who suggested that the donor nuclei from the intestinal cells that actually promoted normal development may have come from primordial germ cells, rather than from intestine. In anuran embryos, the primordial germ cells arise in the endoderm and remain in that location until relatively late in embryonic life. More probably some may be derived from undifferentiated basal cells of the intestine.

The main challenge to the idea that progressive nuclear changes normally occurring during development are responsible for the restriction in capacity to promote normal development of transplant eggs comes from observations that unspecific chromosomal abnormalities frequently occur after nuclear transplantation. Although quite a number of abnormally developing nuclear transplant embryos possess diploid chromosome sets which give a normal appearance, it is difficult to be certain that these developmental abnormalities may not be due to subtle, cytologically undetectable, nuclear changes (chromosome changes) which may have arisen after nuclear transplantation.

Claims that the developmental restrictions following transplantation of nuclei are *ex post facto* results of damage incurred in the transplantation procedure itself are probably not well substantiated. If that were the case, one might want to know what is so unique about blastula nuclei that they suffer so little damage in the experimental procedure, while nuclei of older cells are so much more susceptible to damage. Somatic nuclei from endodermal cells of young donors, although much easier to handle in the transfer procedure, are severely restricted in their capacity to promote normal development, while nuclei obtained from dissociated germ cells, which are much more difficult to handle, give normal development in 40 per cent of test eggs. The suggestion has been advanced that the egg cytoplasm may impose upon the chromosomes of the transplant nuclei from more advanced cell types a division cycle with which they can no longer keep up and that this may lead to chromosomal damage and developmental arrest. Mitotic irregularities might produce chromosome fragmentations or result in deficiencies of DNA replication.

There is evidence that the X chromosome is always late in replicating mammalian cells. There is also evidence that in human blood cells some autosomes replicate late. It is conceivable, therefore, that differentiated nuclei contain some late replicating chromosomes, which may not complete DNA replication in the more rapid mitotic cycle of cleaving eggs. One cannot dismiss lightly the possibility that transplant nuclei may be out of phase with the capacity of the cytoplasm of egg cells to induce rapid division. This possibility presents a real alternative to the doctrine of a built-in programme of nuclear change occurring with development.

Evidence for nuclear-cytoplasmic feedback

In the context of our present concept of gene action in development, the demonstration, provided by the nuclear transplantation experiments, that the cytoplasmic environment provides a 'feedback' on nuclear activity presents us, perhaps, with much more significant clues to differentiation than the question (which the transplantation experiments were initially designed to answer) of whether nuclei undergo stable changes during differentiation. The idea of **differential gene activation** or **repression** during development has been practically elevated to the status of 'central doctrine' of developmental biology. It is, of course, strictly a speculation that is built on the *a priori* assumption that the cytoplasm must interact with the nucleus and alter it in such a way that new gene products can accumulate in the cytoplasm and thus initiate lasting changes among cells that originally were all alike. The nuclear transplantation experiments provided the first systematic evidence that the existence of a nuclear-cytoplasmic feedback is indeed a fact. Nuclear genes do not exist immutably and unchangeably behind the protective cover of the nuclear membrane, like the printer's plates behind glass, but there is a very lively interaction between nucleus and cytoplasmic environment leading to progressive change in the makeup of a cell.

Moore[201] has shown, as mentioned above, that blastula nuclei exposed to the cytoplasm of eggs of a different species are permanently changed, so that they become incapable of promoting normal development even after they have been returned to their own familiar cytoplasm. But the most dramatic evidence for a direct influence of the cytoplasmic environment on the nucleus came from studies by Gurdon and Brown[126]. Nuclei from gastrular and larval cells were transplanted into enucleated eggs of *Xenopus laevis* in order to see whether nuclei obtained from differentiated cells undergo a modification in their pattern of RNA synthesis when transferred to a new and different cytoplasmic environment. An analysis of this kind is possible in *Xenopus* because in amphibians the type of RNA that is synthesized varies greatly with developmental stages. (See Chapter 5, p. 101.) They could show that the nuclei derived from a population of highly differentiated intestinal cells that are actively engaged in ribosomal RNA synthesis, after transplantation into egg cells abruptly stop ribosomal RNA synthesis in the new cytoplasmic environment. Within 40 minutes after successful transplantation, nucleoli disappear from such transplant nuclei.

A similar effect of egg cytoplasm on DNA synthesis has also been reported by Gurdon[125]. In adult frogs, less than 10 per cent of nuclei from mature brain, liver, or red blood cells incorporate tritiated thymidine (an exclusive precursor of DNA) during a 1–2 hour labelling period. If nuclei from these adult cells are transplanted into egg cells that are exposed to ^{3}H thymidine, such transplant nuclei start synthesizing DNA within 90 minutes after having entered the egg cytoplasm. On the other hand, when ^{3}H uridine (an exclusive precursor of RNA) is introduced to eggs that have just received a transplant nucleus from an adult cell no incorporation takes place. Autoradiographs of such egg cells and of nuclei fixed 50 minutes after labelling show no grain count above background. Evidently, egg cytoplasm suppresses ribosomal synthesis very soon after the transplant nucleus is exposed to it.

In a third series reported by Gurdon, nuclei obtained from tailbud and larval tissues that were actively synthesizing ribosomal RNA were transplanted into enucleated eggs and the resulting transplant embryos were examined for the type of RNA synthesis at subsequent blastula stages. The nuclei which were actively synthesizing rRNA at the time of removal from larval cells gave rise, after transplantation, to daughter nuclei, which in the new cytoplasmic environment ceased synthesizing ribosomal RNA and switched to synthesis of DNA, characteristic of normal cleaving cells. Nuclear transplant embryos, whose nuclei were derived from neurula cells, were allowed to develop to neurula stage themselves, whereupon they commenced synthesis of rRNA again just as soon as they began to reach the neurula stage. The onset and rate of rRNA synthesis in such embryos was indistinguishable from control embryos derived from normally fertilized eggs and reared to the same stage.

These observations have since been extended to nuclei from cells of even later stages. Thus, nuclei transplanted from neuroblasts of larval brains into enucleated eggs, switched from rRNA synthesis to DNA-like RNA synthesis or just simply to DNA synthesis. Nuclei obtained from cleaving cells of the morula that normally are mainly engaged in DNA synthesis, when transplanted into oocytes switched to active rRNA synthesis typical of oocyte behaviour. In the oocyte environment the chromosomes of transplant nuclei assumed a configuration characteristic of meiotic nuclei. (For a review, see Gurdon[125].)

Additional observations that can be interpreted in terms of direct intervention by cytoplasmic environment on gene activity during development were recently reported by Briggs and Justus[45] in connection with the gene mutation o in the axolotl. Gene o described by Humphrey[160] exerts a combination of developmental effects on the axolotl, most important of which are those on the germ cells. Females homozygous for gene o when crossed to normal males will give rise to fertilizable eggs, which will arrest at gastrulation. Development of eggs derived from oocytes of o/o females can be enhanced beyond gastrulation and in some cases to larval stages by injection of minute amounts of cytoplasm from normal eggs. The ***corrective cytoplasmic factor*** is found mainly within the germinal vesicle of the oocyte. This cytoplasmic factor is probably a gene product of the normal allele which might exert its corrective action through intervention on nuclear RNA synthesis.

It has been established that in *Xenopus* nuclear RNA synthesis undergoes a rapid and striking increase at the late blastula stage. The stage at which the increase normally takes place coincides with the stage at which the first signs of decline in RNA synthesis begin to appear in mutant eggs obtained from o/o females.

Possibly, embryos of o/o females lack this activation of nuclear RNA synthesis, but minute amounts of cytoplasm of normal eggs containing products of the normal allele might be sufficient to stimulate the relevant genes responsible for this early synthesis.

The nature of this substance has since been investigated. Injection of corrective substance of nuclear sap obtained from normal eggs into the two-cell stage of the mutant promoted normal development only in that portion of the embryo which was derived from the injected blastomere. This has been interpreted to mean that the corrective substance fails to pass through the plasma membrane separating the two blastomeres and must, therefore, be of large molecular weight. This expectation has been borne out in experiments designed to identify the chemical nature and cellular localization of this substance. Cells were subjected to centrifugation and the fraction was isolated and tested. The corrective action remained in the supernatant after all particulates were removed, indicating that the active material seems to be a large molecule that is not associated with any cell organelle. It was furthermore noted that:

1. The material is heat labile and can be completely inactivated if heated to 52°C for 1 hour.
2. Treatment with trypsin at room temperature (20°C) for 1 hour destroys its activity.
3. The proteins present in ammonium sulphate preparations made from normal and mutant eggs have been compared by acrylamide gel electrophoresis. They differ by a single band.

Briggs has not yet isolated the protein of this band in sufficient quantity to test whether it is indeed identical with the component present in normal eggs which is presumably responsible for the promotion of orderly gastrulation and subsequent development.

Another demonstration of cytoplasmic molecules entering nuclei has been provided by **cell fusion** experiments of Bolund et al. (quoted in Gurdon[123]). It has been possible to produce **heterokaryons** consisting of HeLa cells and chicken erythrocytes. Prior to fusion, the erythrocyte nucleus is not engaged in DNA synthesis. After fusion, the erythrocyte nucleus undergoes a seven-fold increase in dry mass (as measured by interference microscopy) presumably due to the entry of cytoplasmic proteins. Concomitant with these changes in weight and volume, the erythrocyte nucleus initiates DNA synthesis.

In another experiment, Arms (quoted in Gurdon[123]) labelled fertilized amphibian eggs with ^{3}H amino acid, and subsequently enucleated them. Two hours after enucleation, when nearly all labelled amino acids had been incorporated into protein, puromycin in concentrations sufficient to repress protein synthesis as well as brain nuclei were introduced into the enucleated eggs. Autoradiography revealed that eggs fixed $\frac{1}{2}$ hour after nuclear transplantation had been completed contained labelled protein in their transplanted nuclei, and that the concentration of labelled material was twice that of the cytoplasm. It is concluded that the protein that was recovered in the nucleus must have been transferred from the cytoplasm.

Merriam (quoted in Gurdon[123]) also introduced adult brain nuclei into eggs whose cytoplasmic proteins had been previously labelled by supplying ^{3}H amino acid during oogenesis. Such brain nuclei enlarged after transplantation into the egg cells, and within a short time began to concentrate the labelled protein. A few of the nuclei that did not enlarge after transplantation failed to accumulate the labelled protein. Further-more, the only nuclei that responded to transplantation to the new environment of the egg cytoplasm by switching from RNA to DNA synthesis were the ones that had also taken up cytoplasmic protein. Apparently the entry of cytoplasmic protein into nuclei is causally related to the change imposed upon nuclear activity by the cytoplasmic en-vironment.

DIFFERENTIAL RELEASE OF GENETIC INFORMATION

Chromosomal puffing and gene action in Diptera

The nature of the polytene chromosomes

Of the possible control mechanisms that might bring about a situation whereby cells that are orginally of identical genetic constitution might acquire different protein patterns, the most efficient one that comes to mind would be control at the site where reading of the genetic informa-tion takes place. Stimulated by the demonstration that in bacteria differential transcription is responsible for adaptive enzyme induction, biologists have become partial to the idea that transcription of messenger molecules, through differential gene activation, initiates the arising of differences in embryonic cell populations.

Fortunately, nature has helped the embryologist by giving him **giant**, or **polytene**, **chromosomes** to work with, for the discovery of these in some tissues of dipteran flies has made genes 'almost' visible to the trained eye, if supported by the resolving power of an immersion lens. Giant chromosomes, which exist in these creatures in a variety of tissues such as gut, Malpighian tubules and the salivary gland, are bundles consisting of 500 to 1000 chromosomal threads existing in a state of despiralization. The threads, or **chromonemata,** stick together in such a way that identical sites or chromomeres lie side by side, forming discs separated by regions of nonchromonematic material. These polytene chromosomes may attain a length of ten times that of normal univalent chromosomes. This unusual chromosomal growth is probably related to the circumstance that the salivary gland and certain other tissues of dipteran larvae grow mainly by increase in cell size, rather than by cell duplication. Assuming that there exists a critical ratio between a given cell volume and the amount of chromosomal material required to maintain it in the absence of cell division, repeated chromosomal reduplication would be necessary to support a cell whose volume increases excessively.

The lucky circumstance to the biologist lies in the fact that because of the increased diameter of such giant chromosomes, more and more detail of the fine structure of the chromosomes can be revealed. The polytene chromosomes thus provide the closest approach to an actual image of the gene.

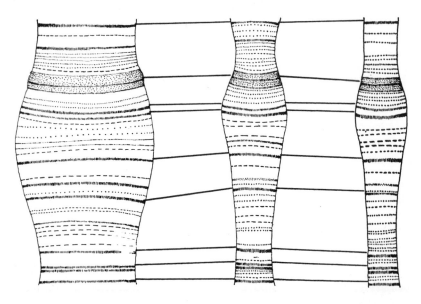

Fig. 6.4 Giant chromosome number III of larva of the midge *Chironomus* varies in shape and slightly in banding pattern depending on the type of cell in which it is found. Identical sections here are from the salivary gland (left), Malpighian tubule (middle) and rectum (right). (From Fischberg and Blackler[88])

The significance of chromosomal puffs

The most exciting aspect to the embryologist concerns the observation that at times portions of these chromosomes swell up into **puffs** and that the appearance of these puffs follows a specific pattern of occurrence, both with respect to developmental age, as well as to cell type. Within a given chromosome, the sites at which puffs appear in young larvae differ from those of older and more mature ones. A comparison of such diverse tissues as salivary gland, Malpighian tubule and rectum, furthermore, reveals that homologous chromosomes in these three cell types swell and give rise to puffs at very distinct regions. It has been hinted by Beerman[19-22] that the differences in **puffing patterns** might represent the long sought and hoped for visualization of differential gene

messenger RNA contributes either wholl[
the chromosomes.

The easiest proof would be to isolate th[
then to test it for template activity. This t[
Braverman, Gold and Eisenstadt (quote[
RNA fraction which also had an A/U rat[
(reminiscent of puff RNA), which did hav[

Induction of puffing pattern by transplanta[
different environments

In 1962, Kroeger (quoted in Fischberg[
ingenious experiment. Salivary gland nu[
transplanted into egg cytoplasm of differe[
effort to induce differential chromosom[
selected: (1) **Preblastoderm** eggs, i.e. e[
tion, and (2) **Blastoderm** eggs, i.e. eggs[
formation. The procedure consisted of pl[
gaster onto a glass slide covered with oil. T[
their contents squeezed onto the slides an[
form one big droplet. Larvae of *D. busckii*[
removed for nuclear explantation and pla[
to the egg cytoplasm. Nuclei were dissect[
pushed through the oil into the droplets o[
for three hours. Controls consisted of[
incubated in pure oil droplets. When sali[
transplanted to preblastoderm and blas[
appeared at chromomeric regions that ha[
the salivary cells before, while other puffs[
chromosomes rapidly disappeared. (See[
the gland, which in the prepupal and pupa[
of any puffs, was induced to swell at cert[
after it was placed into the preblastode[
appears only during the preblastoderm pe[
might represent a product of a gene that i[
of embryonic development, possibly prom[
of yolk. Such a gene would not be expecte[
nuclei at all but might be induced to acti[
the preblastoderm cytoplasm. Other puffs[
products that are required in the product[
example, saliva. There are puffs known th[
nuclei of prepuparian stages which disapp[
formation when production of saliva is n[
of this is provided by the X chromosome[

activation. This idea has been reinforced by the demonstration that these puffs are really sites of RNA synthesis and that a good proportion of the RNA synthesized at the puffs is messenger RNA[19-22]. When tritiated uridine, a precursor of RNA, was injected into larvae, radioautographs of salivary glands prepared within minutes after injection revealed appreciable amounts of label within the nuclei, but none in the cytoplasm of the salivary gland. The labelled RNA was located almost exclusively in the puffed region of the chromosomes and in the nucleolus. After prolonged incubation time, the cytoplasmic RNA also became labelled, suggesting a transfer of labelled material from chromosomal puffs to cytoplasm. The portion of the chromosomes not engaged in forming puffs remained unlabelled[19-22]. Autoradiographs of salivary glands of larvae injected with tritiated thymidine, a precursor of DNA synthesis, showed that only two out of fifty cells of the salivary gland had incorporated the label, a figure roughly corresponding to the number of cells presumably in the replication phase at any one time.

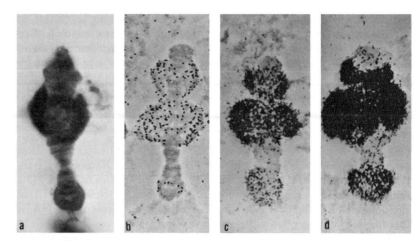

Fig. 6.5 The small chromosome IV from the salivary gland of *Chironomus tentans*. (**a**) Stained for RNA and DNA with toluidine blue (note intense RNA staining in puffs). (**b**), (**c**), and (**d**) Radioautographs after incubation of living glands in ³H-uridine showing different overall labelling as well as relative differences in the uptake of label between puffs of different size. (From Beerman[21])

Evidence that chromosomal puffs are sites of mRNA synthesis

In order to establish that chromosomal puffs are really sites of differential gene activation, it is necessary to characterize the RNA produced in these puffs and to show that the puffing pattern of the different chromosomes varies during the development of cell types.

0·01–0·3 per cent of the total length
hybridizes much more easily than tł

As was pointed out above, mRN
ribosomal rRNA by its overall base
DNA, containing 40 per cent G + C ;
Chironomus is of the A/T type with ;
70 per cent and a G/C composition
RNA fractions from salivary gland (
the puffs, has a G + C content lower
cannot be satisfactorily explained ;
chromosomal RNA to be about 30 p
of DNA.

A more elegant attempt to recog
duced in the puffs was made by Be
strom, through determination of ba
requires as a first step extraction of R
RNAase digestion. The fractions ob
tion are then subjected to hydrolys
phoresis on rayon fibres of 25 μm di
identified by UV absorption. The rs
uracil (U), cytosine (C) and guanine (
nucleolar and nuclear) may then be (
cent). The underlying assumption is
the nucleus (in this case, from the pu
in base ratio from nucleolar RNA or
(ribosomal), it is not very likely that t
precursor of the latter (nucleolar ar
that the base ratios of RNA obtainec
as of chromosome IV were radically
from nucleolar and cytoplasmic
to be double stranded with base (
mentary to the other, so that the
and guanine to cytosine (G/C) shoulc
from chromosomal puffs of some o
to be extremely asymmetrical, with
of G/C and A/U. Upon closer ex
noted that values for adenine w
values were also considerably hi
explain this observation, Beerman
at the puffs may be either single
RNA strand is produced, or that
are made, one is immediately de
All these findings thus strengthen

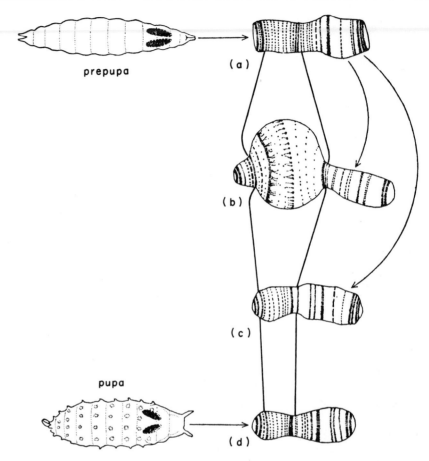

Fig. 6.6. Experiment performed by H. Kroeger involved chromosome number II from the salivary gland of the fruit fly *Drosophila busckii*. The chromosome is depicted here as it appears at (**a**) the prepupal and (**d**) the pupal stages of the fly's development; and as it appeared after transfer to (**b**) the preblastoderm, and (**c**) the blastoderm egg contents of *D. melanogaster*. (From Fischberg and Blackler[88])

Transplantation of the X chromosomes from nuclei of actively secreting salivary gland cells into preblastoderm and blastoderm cytoplasm leads to rapid disappearance of some of their puffs. This would be expected, since genes necessary for salivary gland function would not be active in egg cytoplasm. These observations can be interpreted to mean that in the developmental sequence, different chromosomal segments are mobilized to elaborate their product, and that a changing cytoplasm feeds back on the nucleus to trigger further differential gene activation. The only comparable system where the process of chromosomal RNA synthesis has also been made visible at the level of the ordinary microscope is the lampbrush chromosome of amphibian oocytes described by Gall[91] and Gall and Callan[92].

A word of caution is, however, in order. The rationale for the argument that differential puffing is intimately related to cell differentiation is based on the observation that the puffing pattern varies characteristically from tissue to tissue and from stage to stage. There are some objections to this generalization:

1. There are only very few puffs which so far have been found to be present exclusively in one tissue or only in cells with identical function.

2. Most puffs are found in more than one tissue, and many are present in all tissues, though they frequently do not occur in all tissues of the same individual at a given time. It may be that most of the genes represented by these puffs are active in more general functions of cell metabolism which are not specific to only one type of cell but are shared by many or all cells (for example, genes coding for respiratory enzymes).

3. Loci that are not in a puffed condition in the salivary gland of one larva may be puffed in the gland of another. The more larvae one examines, the fewer puffs are left that are specifically and exclusively present only in one cell type.

4. In certain regions, but not in all, one can observe the appearance of the same puffs after relatively nonspecific treatments like temperature shocks and exposure to 1·2 per cent and 1·4 per cent KCl solution. It seems likely that regions that are able to react indiscriminately to changes in the extranuclear milieu may represent genes whose products need to be amplified many times and are, therefore, most labile to extranuclear influences. The majority of genes probably transcribe their templates in more orthodox fashion and, because they do not accumulate their product, may never show puffs.

Triggering the puffs: the role of hormones

The observation that during development from larval to prepupa to pupa a sequence of puffing patterns occurs at a given site, alternating with the disappearance and appearance of puffs at new and different sites along the axis of the chromosome, started developmental biologists on a search for inducers that might trigger 'genes' into the kind of synthetic activity so obligingly made visible in the giant chromosomes.

This search was rewarded when in 1961 Clever[60] found that the injection of trace amounts of the moulting hormone, ecdysone, into *Chironomus*

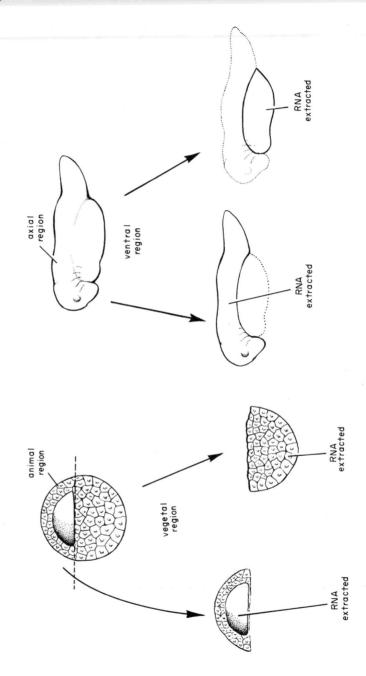

2. DNA was obtained from blood cells of adult *Rana pipiens* and precipitated with cold ethanol.

3. RNA was collected from animal, vegetal, axial and ventral regions of blastulae and tailbud stages respectively.

4. The hybridization procedure that was used took advantage of the fact that single stranded DNA can be trapped on nitro-cellulose filters where it can hybridize with RNA possessing complementary base sequence. Aliquots of heat denatured frog lymphocyte DNA were trapped on such nitro-cellulose filters. Increasing amounts of ^{3}H RNA were then added to the vials containing filters on which denatured single stranded DNA had been trapped until saturation levels were obtained so that no more RNA could combine with the DNA on the filter. Following 12 hours of incubation, the filters were washed, treated with RNAase, and prepared for counting in scintillation counters for 10 minute periods. It was found that the denatured DNA on the filters could be saturated with 1 mg of ^{3}H RNA isolated from animal halves. Further addition of animal RNA did not anneal to the DNA as indicated by the failure of radioactivity to increase in the system. Addition of ^{3}H RNA from vegetal halves cultured for 24 hours in ^{3}H uridine did, however, anneal to the DNA previously saturated with RNA from animal cells, as indicated by the sudden increase of radioactivity. Reciprocal experiments revealed additional annealing when RNA isolated from animal halves of the eggs was added to filters that had been saturated with ^{3}H RNA from vegetal halves. Similar results were obtained by experiments using dorsal and ventral halves of tailbud stages.

Flickinger *et al.*[90] interpreted the results of their experiments to mean that different populations of messengers are produced in the animal and vegetal and axial and ventral regions of the amphibian egg and embryo, respectively. On the other hand, their data also revealed that a major portion of messenger species is held in common by the two different regions (animal–vegetal, axial–ventral).

Competition experiments

Another very instructive approach for the demonstration of differential transcription is provided by the competition experiments of Denis[78].

Hybridization competition experiments are based on the rationale

---

Fig. 6.8 Diagrammatic representation of experiment by Flickinger, Greene, Khol and Myagi.[90] RNA from animal and vegetal pole and from axial and ventral region was extracted and annealed to frog lymphocyte DNA. Hybridization of RNA from these four regions with single-stranded DNA indicated that each area contained some messenger RNA that was unique to each region tested.

that two molecules of RNA which have similar base sequences compete with each other for the same or complementary site on the DNA, whereas unrelated RNA molecules do not compete.

In these experiments a fixed amount of DNA is incubated with a given amount of labelled RNA which has been extracted from an embryo at a given stage of development. The labelled RNA obtained from the embryo that has been exposed to a brief pulse of radioactive precursor is called the *reference RNA*. To this mixture is then added an increasing amount of nonradioactive RNA obtained from the same or earlier or later embryonic stages. This is the *competitor RNA*. If the reference RNA and the competitor RNA are identical, then the latter (competitor RNA) will progressively replace the former (reference RNA) and a decrease in the percent of DNA binding or hybridization of labelled (reference) RNA will ensue. Since the hybridization of labelled reference RNA is measured by the amount of radioactivity in the system, any decrease in radioactivity will be an indicator of the extent of competition exerted by the competitor RNA. If the labelled reference RNA and the cold competitor RNA are from the same source or are identical the latter will progressively replace the former or dilute the radioactive RNA. If the two RNAs under comparison have no nucleotide sequence in common, the addition of increasing amounts of competitor RNA will have no influence on the hybridization of the (labelled) reference RNA.

Using this method, Denis[78] addressed himself to the question whether all the messenger molecules that exist in a given embryonic stage are present already at earlier stages or are still present at later stages, or whether specific messenger molecules appear briefly during embryogenesis and never reappear again.

Figure 6.9 shows results of just such an experiment. A tadpole at swimming stage (Stage 42) was given a pulse label and the labelled RNA extracted from this source served as reference RNA. Cold RNA (unlabelled RNA) extracted from cleaving frog eggs, gastrulae, neurulae and tailbuds served as competitors. It can be seen from Fig. 6.9 that RNA from cleaving eggs does not interfere or compete at all with hybridization between DNA and RNA extracted from Stage 42. On the other hand, RNA from gastrulae and from later embryos increasingly competes with RNA of cells from differentiated tadpoles (Stage 42). Therefore it can be concluded that none of the nucleotide sequences that are labelled by a short pulse in differentiated cells of Stage 42 exist already in cleaving eggs. On the other hand, an increasing proportion of messengers that are present at Stage 42 are already present among the RNA population present in gastrulae and neurulae stages. Conversely, some messengers that appear in the gastrula are not yet present in cleaving

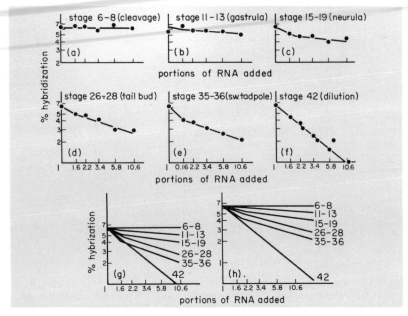

Fig. 6.9 Competition experiment between pulse-labelled RNA from Stage 42 (differentiated tadpoles) and nonlabelled RNA from earlier embryos. In (**g**) the results of (**a**) to (**f**) are summarized. In this diagram the stage-to-stage comparison is based on the addition of nonlabelled RNA from different stages. Since the RNA content of the embryo approximately doubles between fertilization and the uptake of food (4 μg in the egg; 8 μg in the differentiated tadpole), 4 μg of egg RNA contain the mRNA from one embryo, whereas 4 μg of tadpole RNA contain the mRNA of only one-half embryo. To compare the competing power of the mRNA extracted from one embryo of various stages, the abscissa scale of (**g**) has to be changed. In (**h**) are shown the same results as in (**g**), but plotted on a per embryo basis. (From Denis[78])

egg cells, and a small proportion of mRNA produced by gastrulae and neurulae stages is no longer present in cells of later embryos.

A similar experiment was reported by Church[59], who compared the transcriptional pattern of differentiated mouse tissues and mouse blastocysts. In his experiment, RNA was synthesized *in vitro* by mouse liver and by mouse kidney chromatin and the labelled newly synthesized RNA (reference) was then hybridized to mouse DNA. Cold unlabelled RNA was extracted from brain, kidney and liver and was used as competitor. It can be seen from Fig. 6.10 that messenger RNA extracted from liver cells is the most efficient competitor with RNA transcribed from liver chromatin (o per cent hybridization), while kidney and brain RNA competed to a much lesser extent with the liver reference RNA, and

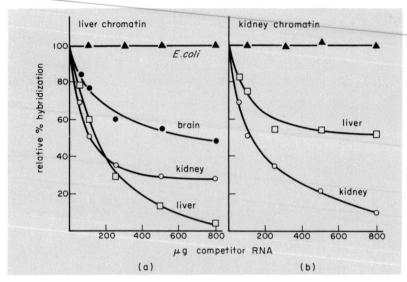

Fig. 6.10 Tissue specificity of RNA synthesized from chromatin originating from liver or kidney using mouse RNA polymerase. (a) RNA was synthesized *in vitro* from a mouse-liver chromatin template using mouse RNA polymerase. The reaction was run in 5 ml buffer at 37°C for 10 minutes, and the RNA extracted with hot phenol. RNA isolated from various tissues was used as competitor for the hybridization of the *in vitro* synthesized RNA to DNA. Hybridization with 12 μg DNA in 0·2 ml of 2× SSC at 67°C for 18 hours. 384 cpm were hybridized in the absence of competitor. (b) RNA was synthesized *in vitro* from a mouse-liver chromatin template using mouse RNA polymerase. The reaction was run in 5 ml buffer at 37°C for 10 minutes and the RNA extracted with hot phenol. RNA isolated from various tissues was used as competitor for the hybridization of the *in vitro* synthesized RNA to DNA. Hybridization was performed with 12 μg of DNA in 0·2 ml of 2× SSC at 67°C for 18 hours. 336 cpm were hybridized in the absence of competitor. (From Church[59])

bacterial RNA did not compete at all. (The level of hybridization was kept at 100 per cent). Similar results can be seen when kidney chromatin served as a template for reference RNA.

In Fig. 6.11 competition was measured between RNA obtained from mouse blastocysts (serving as reference RNA) and RNA from such diverse sources as 9·5-day mouse neural tissue, chick embryos and older mouse embryos. The 100 per cent competition by RNA of 10-day-old whole mouse embryos with reference RNA extracted from mouse blastocyst suggests that the whole embryo contains all the RNA transcripts that are present in blastocyst cells already. On the other hand, the RNA from cells of the 9·5-day neural tissue does not contain all the RNA sequences that are present in the blastocyst. This again indicates

that in the blastocyst some RNA is made which subsequently disappears. Whether neural tissue contains new species of RNA which have not yet been made in the blastocyst has yet to be determined. We merely know that blastocyst cells contain RNA transcripts that are no longer found in neural tissue. The partial competition by the whole 7-day chick embryo suggests a great deal of conservatism in developmental genes during evolution. Bacterial RNA on the other hand is so foreign, that competition between it and mouse blastocyst RNA is totally absent.

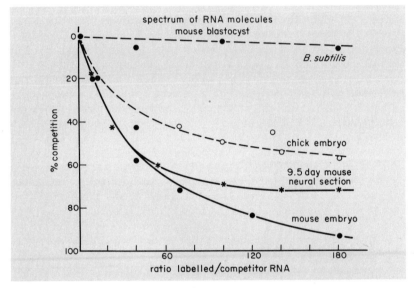

Fig. 6.11 The competition of unlabelled RNA isolated from whole 7-day chick embryos, whole 10-day mouse embryos and 9·5-day mouse neural tissue in the reaction of labelled blastocyst RNA with unique mouse DNA sequences. Reactions were incubated with formamide under conditions equal to DNA C_{ot} of 10 000. (From Church[59])

Implication of selective gene action for theories of differentiation

Excess DNA

A logical corollary to the theory of selective gene action during growth and differentiation is the proposition that the nucleus of a cell must be inhabited by large amounts of genetic material inert at any one time. The presence of ***inactive DNA*** that may never be involved in transcription of any messenger molecules during a given stage of differentiation was first suggested by Mirsky[196] and more recently by Davidson[75]. Mirsky, Alfrey and Davidson had discovered some time ago that

nuclei isolated from calf lymphocytes can indeed survive in suitable medium and still be metabolically quite active. In their system such isolated nuclei continued to synthesize not only protein but also DNA and RNA, the RNA presumably being mainly of the messenger type because its synthesis could be prevented by actinomycin D. Treatment of such isolated nuclei with DNAase removed 80–85 per cent of the DNA present. Removal of the remaining 15–20 per cent of DNA required treatment of nuclei with considerably larger enzyme concentrations and longer incubation time. The important finding revealed by this study was that 60 per cent of the nuclear DNA could be removed without impairing RNA synthesis. Mirsky[196] reasoned that if nuclear RNA synthesis can continue even after so much of its DNA is destroyed, it would seem that the 'easily' removable DNA fraction represents the inactive portion of the genetic material of the nucleus, while the remainder represents the **genetically active fraction.** This argument could be strengthened if, in fact, it could be shown that 15–20 per cent of the nuclear DNA remaining after DNAase treatment is required to ensure continued RNA synthesis.

Masking and unmasking of DNA

The idea that only a few genes in an organism actually transcribe, suggests, of course, as a corollary that much of the DNA might be 'masked' by some unspecific substance. There is good evidence that template activity of DNA is restricted in preparations of chromatin.

In an *in vitro* system containing as primer either *free DNA* or *chromatin-bound DNA* it was possible to follow the rate of synthesis of RNA. In such a system equivalent amounts of DNA gave rise to very different rates of RNA synthesis, depending upon whether free DNA or DNA bound to chromatin is used as primer. With free DNA the synthetic reaction proceeds at a much higher rate than with chromatin-bound DNA. This raises the question of whether the low yield of RNA that is synthesized when chromatin-bound DNA serves as a template is due to a general kinetic inhibition under these conditions or whether a greater variety of DNA sites are transcribing when the molecule is not bound to chromatin. The answer was given by the following experiment.

In the procedure illustrated by Fig. 6.12, RNA was synthesized *in vitro* using either mouse liver chromatin or purified, free mouse liver DNA as templates. The labelled newly synthesized RNA was then hybridized to mouse DNA. As can be seen from Fig. 6.12, the proportion of hybridization is much greater with RNA obtained when free DNA served as a template than with RNA obtained when chromatin-bound DNA was transcribing. Far more of the mRNA obtained from free DNA hybridized than did messengers obtained from chromatin bound

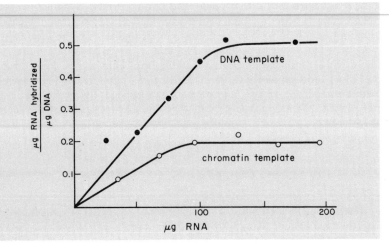

Fig. 6.12 Saturation of DNA and RNA synthesized *in vitro* using liver chromatin of DNA isolated from liver chromatin as primer. RNA was synthesized in 50 ml of the standard reaction mixture using 50 μmoles of either liver chromatin or DNA isolated from liver chromatin. The reaction was run for 10 minutes and then the RNA extracted with hot phenol. Hybridization was carried out using increasing amounts of RNA with 20 μg of DNA in 0·2 ml of 4× SSC at 65°C for 18 hours. Specific activity of RNA synthesized on a DNA template was 810 cpm/μg and 720 cpm/μg for RNA synthesized from chromatin. (From Church[59])

DNA. Apparently there is a greater variety of messengers produced from free DNA than from chromatin-bound DNA. It might therefore be assumed that certain components of the chromatin act as a block to DNA-dependent transcription. This is in line with the well-established generalization that the **histones** which make up part of the protein framework of chromosomes repress DNA transcription.

Evidence has accumulated in recent years supporting the proposal that complexing DNA with histones suppresses the activity of genes. Primarily this evidence is derived from experiments showing increased RNA synthesis upon removal of histone from nuclear preparations, and from the demonstration that DNA-histone complexes are incapable of 'priming' RNA synthesis *in vitro* in preparations of pea seedling nuclei. Conversely, greatly increased RNA synthesis has been reported following deproteinization of such chromatin preparations.

On the basis of the inverse relationship between histone content and priming activity, and in accordance with the generally held view that genes are mostly inactive until the time of gastrulation, one would expect the chromatin of the earlier stages to have a higher histone content than the latter. However, this does not seem to be the case. As a matter of fact, it has been found that nuclei and chromosomes in mouse, frog and snail embryos begin to stain strongly for histones at the time of gastrulation but not before. This seeming contradiction may be resolved by assuming that many genes are in fact active before gastrulation, but that the mRNA produced is not translatable until later stages. (See Chapter 5.) Thus the frequently cited evidence that there are no demonstrable paternal effects in embryos until gastrulation need not imply that genes are turned off until that time, but that the absence of paternal effects is due to the presence of previously made but masked mRNA species which cannot yet be translated.

SOME FINAL SPECULATIONS ON GENES IN DEVELOP-MENT

The evidence for feedback of cytoplasm on nuclei, so carefully put together by Briggs, King, Moore, Fischberg, Gurdon and Kroeger, is overwhelming and from it we have learned that nuclei behave and synthesize in accordance with the cytoplasm in which they find themselves. The most plausible assumption advanced so far is the idea that once very subtle cytoplasmic differences are established between two or more cells, differential gene activation will initiate the synthesis of distinct species of molecules and thus compound the differences in geometric progression.

Some control of differentiation must, however, almost certainly be exerted on the level of translation, for we know that a good deal of unstable RNA is rapidly synthesized and degraded in the nucleus and never reaches the cytoplasm, and therefore can never be translated. We also know from hybridization studies that nuclear RNA hybridizes far more efficiently with DNA than does cytoplasmic RNA. This suggests that translation of genetic information is a two step process: (1) There must be differential gene transcription and (2) a selection must be made as to which of the transcribed products shall be transported into the cytoplasm for translation and which of the products shall be degraded in the nucleus.

Lastly, the concept that DNA sites code only for assembly of specific proteins is probably much too simplistic for an understanding of the manner by which genes control differentiation.

One of the most perplexing discoveries in molecular biology in recent years has been the fact that each cell of a vertebrate contains far more DNA than is required to code for all its proteins. Mammals contain about 7 picograms of DNA per cell—about 1000 times as much as is required to code for all their proteins. What can be the role of this excess DNA?

The excess DNA for which no specific function has yet been dis-

covered is assumed by many to be inert, or it is sometimes assigned a regulatory role in the quantitative functioning of the structural genes. In fact, certain identifiable genes do regulate the function of other genes, but their number is small and theoretically need not be larger than the number of structural genes. Perhaps most of the DNA simply regulates in some as yet unknown fashion the quantitative function of the structural genes. The vast array of genetic differences among individuals are primarily differences in quantity not quality. The majority of genes may not transcribe specific proteins at all, but may be merely involved in a very finely graded control of the relative activities of the structural genes at various stages in cell differentiation.

According to the 'central doctrine' of experimental biology, differentiation reflects the emergence of selective protein synthesis, an event which is presumably preceded by differential repression or derepression of genes that code for these proteins. The demonstration that differential gene activation parallels the more visible changes encountered during the development of the embryo is indeed exciting, but it does not end our search. Control of development at the genetic level, whether exerted on transcription or translation of the message, implies that the differences that are introduced into populations of embryonic cells are consequences of changes in the biochemical household, i.e. the stock of proteins of these cells. The next question that must be raised by the developmental biologist is 'Protein synthesis for what?' Are the new proteins or the RNA codes that initiate their synthesis mobilized for mitotic proliferation? Are they required for cell death and lysis of embryonic cells that must be removed, or are they required for stimulation of migration patterns, or for surface reactions leading to cell recognition and aggregation, or for any of the other processes, like induction, that are operating in the modelling of cells into tissues, organs and structures?

If development is essentially a sequential spelling out of genetic messages, and if the genetic sentences can only spell out proteins or related macromolecular substances, then it follows that the embryo must contain a host of short-lived morphogenetic substances that in turn stimulate such events as induction, proliferation, differential cell death, cell migration and cell recognition. All of these morphogenetic events have been shown to be under genetic control. (See Chapter 5.) The only other alternative would be to assume that genes can direct development in yet some other way than by merely acting as codes for protein synthesis.

We are rightly proud, thanks to the giants in the field, that we can now distinguish by weight at least four classes of functional RNA molecules, and the order of their appearance in the embryo. But knowing this is only surface knowledge, comparable in scope to the insight a traveller might have acquired if all he had learned from his trip was that the inhabitants of the country he visited consist of underweight, overweight and medium weight people, and that each of these categories of men have different occupational preferences.

Whatever the answer that we can give to our questions, it is well to remember that our present knowledge about gene action in development marks only the beginning.

Epilogue

It is always a little presumptuous to be a prophet in science and predict the direction toward which future research will or ought to turn. Such predictions usually turn out as successful as forecasting next week's weather. However, the temptation to restate once more the major generalizations obtained from the studies and experiments that have been presented in the preceding pages, and the major unsolved problems that remain, is too great to be passed by.

Developmental biologists in the last 15 years have addressed themselves increasingly to solving the logical dilemma presented by our concepts of the genetic control of differentiation. The logical dilemma can be stated like this (see also Chapter 6, p. 108): If visible differentiation of cells merely reflects the emerging differences in their protein and enzyme content, and if the code for all proteins is spelled out by genetic messages reaching the cytoplasm from the nucleus, then how can cells ever become different if all cells contain the same genes?

To resolve this dilemma, an increasing number of investigators have supplied evidence in support of the proposition that although all cells contain all the instructions for making all the proteins found in an organism, not all of the messages inscribed in the nucleus are actually transmitted and sent out into the cytoplasm.

The discovery of chromosomal puffing in Diptera, the work on lamp-brush chromosomes in amphibian oocytes, and the DNA-RNA hybridization studies in amphibian gastrulae, lend strong support to the contention that differential gene activation occurs in development. Studies of transplantation of amphibian nuclei from cells of different stages of development into egg cells certainly provide powerful evidence that the type of genetic messages that are transmitted is influenced by the cytoplasmic environment in which the nucleus is placed.

There remains the troubling question (somewhat of the 'Which comes first, the chicken or the egg?' variety) of how the initial cytoplasmic differences are established. A case can be made for the assumption that there is built into the process of cleavage a mechanism for the emergence of subtle cytoplasmic differences. A two-cell stage may appear to represent two identical cells, but are they really completely alike? In the *Fucus* egg, for instance, the first cleavage already introduces differences in position between the two cells *vis-à-vis* their environment which starts one cell on the path to form rhizoids and the other to form a stem. Once the most subtle cytoplasmic difference is initiated, the process is self-perpetuating through differential gene activation which will lead to elaboration of different gene products in the cytoplasms of different cells, the accumulation of which, in turn, will continue to feed back on nuclear genes.

Thus, as we have seen in the preceding chapter, the embryo must contain a host of short-lived morphogenetic substances. The identification of these, as well as the spelling of the corresponding sentences according to which they are put together in the embryo, remain problems for the future, yet the hope of many developmental biologists, that DNA and RNA will provide the key that will open most or all the doors separating them from the mystery of development, has probably been strengthened by the cumulative results of the last 25 years of experimentation.

In the economy of nature, the same combination of chemical letters that are used to spell out the most involved biochemical instructions, are apparently also employed to signal embryonic cells into action, i.e. to act as inductors. The discovery of an exchange of signals between embryonic cells, first described by Spemann as induction, has since been confirmed for numerous interacting systems as an indispensable requirement to start processes of differentiation. So impressed were embryologists by this phenomenon that for a long time the problem of the nature of the organizer dominated their thinking. We are now confident that we know the identity of the inductors in a great many interacting systems, but the events responsible for the transformation of the 'competent' tissue when 'touched' by the inductor substance are still largely unknown and remain to be explored by investigators of the future.

Lastly, development is more than the emergence of differences in cells, and it does not stop when histogenesis is completed. Development is a construction job of the first magnitude. Processes of growth, cell migration, cell recognition, cell aggregation and disaggregation, selective cell death, to name just a few, must obviously be mobilized to mould organs and structures. There are sophisticated descriptions of these processes to be found in the embryological literature, but the analysis of their mechanisms invites the attention of future investigators to design experiments to test new hypotheses or to formulate theories that will stimulate new experiments.

Of such is the making of embryologists

References

1. AMPRINO, R. (1965). Aspects of limb morphogenesis in the chicken. In *Organogenesis*, ed. DEHANN, R. L., and URSPRUNG, R., pp. 255–81. Holt, Rinehart and Winston, New York.
2. AUERBACH, R. (1954). Analysis of the developmental effects of a lethal mutation in the house mouse. *J. exp. Zool.* **127**, 305–30.
3. AUERBACH, R. (1960). Morphogenetic interactions in the development of the mouse thymus gland. *Devl Biol.*, **2**, 271–84.
4. AUSTIN, C. R. (1948). Function of hyaluronidase in fertilization. *Nature, Lond.* **162**, 63–64.
5. AUSTIN, C. R. (1961). *The Mammalian Egg.* Blackwell, Oxford.
6. AUSTIN, C. R. (1965). *Fertilization.* Prentice-Hall, Englewood Cliffs, New Jersey.
7. AUSTIN, C. R. (1967). *Ultrastructure of Fertilization.* Holt, Rinehart and Winston, New York.
8. AUSTIN, C. R. and BISHOP, M. W. H. (1957). Preliminaries to fertilization in mammals. In *The Beginnings of Embryonic Development*, eds TYLER, A., VON BORSTEL, R. C. and METZ, C. B., pp. 71–107. Am. Ass. Advance. Sci., Washington, D.C.
9. AVE, K. KAWAKAMI, I. and SAMESHIMA, M. (1968). Studies on the heterogeneity of cell populations in amphibian presumptive epidermis with reference to primary induction. *Devl Biol.*, **17**, 617–26.
10. BALINSKY, B. I. (1965). *An Introduction to Embryology*, 2nd ed. Saunders, Philadelphia.
11. BARTH, L. G. (1939). Oxygen consumption of the parts of amphibian gastrula. *Proc. Soc. exp. Biol. Med.*, New York, **42**, 744–46.
12. BARTH, L. G. (1941). Neural differentiation without organizer. *J. exp. Zool.*, **87**, 371–83.
13. BARTH, L. G. (1942). Regional differences in oxygen consumption of the amphibian gastrula. *Physiol. Zool.*, **15**, 30–46.

14. BARTH, L. G. (1946). Studies of the metabolism of development. The metabolic nature of the block to gastrulation in hybrid eggs. *J. exp. Zool.*, **103**, 463–86.

15. BARTH, L. G. (1953). *Embryology*. rev. ed., Dryden Press, New York.

16. BARTH, L. G. and BARTH, L. J. (1954). *The Energetics of Development*. Columbia University Press, New York.

17. BARTH, L. J. (1964). *Development, Selected Topics*. Addison-Wesley, Reading, Massachusetts.

18. BAUTZMANN, H., HOLTFRETER, J., SPEMANN, H. und MANGOLD, O. (1932). Versuche zur Analyse der Induktionsmittel in der Embryonalentwicklung. *Naturwissenschaften*, **20**, 971–74.

19. BEERMAN, W., (1959). Chromosomal differentiation in insects. In *Developmental Cytology*, ed. RUDNICK, D., pp. 83–103. Ronald, New York.

20. BEERMAN, W. (1963). Cytological aspects of information transfer in cellular differentiation. *Am. Zool.*, **3**, 23–32.

21. BEERMAN, W. (1964). Control of differentiation at the chromosomal level. *J. exp. Zool.*, **157**, 49–62.

22. BEERMAN, W. (1966). Differentiation at the level of the chromosome. In *Cell Differentiation and Morphogenesis*. Int. Lecture Course, Wageningen, The Netherlands, pp. 24–54. North Holland Publishing Co., Amsterdam.

23. BENNETT, D. (1963). Embryological effects of lethal alleles in the t region. *Science, N.Y.* **144**, 263–67.

24. BERG, W. E. (1950). Lytic effects of sperm extracts on the eggs of *Mytilus edulis. Biol. Bull. mar. biol. Lab., Woods Hole*, **98**, 128–38.

25. BONNER, J. T. (1952). *Morphogenesis: An Essay on Development*. Princeton University Press, Princeton, New Jersey.

26. BONNEVIE, K. and BRODAL, A. (1946). Hereditary hydrocephalus in the house mouse. IV. The development of cerebellar anomalies during foetal life with notes on the normal development of the mouse cerebellum. *Naturv. Kl.*, **4**, 1–60.

27. BOREI, H. (1948). Respiration of oocytes, unfertilized eggs and fertilized eggs from *Psammechinus* and *Asterias. Biol. Bul. mar. biol. Lab., Woods Hole*, **95**, 124–50.

28. BOREI, H. (1949). Independence of post-fertilization respiration in the sea urchin egg from the level of respiration before fertilization. *Biol. Bull. mar. biol. Lab., Woods Hole*, **96**, 117–22.

29. BRACHET, J. (1938). The oxygen consumption of artificially activated and fertilized *Chaetopterus* eggs. *Biol. Bull. mar. biol. Lab., Woods Hole*, **74**, 93–98.

30. BRACHET, J. (1941). La localisation des acides pentosenucléiques dans les tissues animaux et les oeufs d'Amphibiens en voie de développement. *Archs Biol. Paris*, **53**, 207–57.

31. BRACHET, J. (1947). The metabolism of nucleic acid during embryonic development. *Cold Spring Harb. Symp. quant. Biol.*, **12**, 18–27.

32. BRACHET, J. (1947). Nucleic acids in the cell and the embryo. *Symp. Soc. exp. Biol.*, **1**, 207–24.

33. BRACHET, J. (1948). Le rôle et la localisation des acides nucléiques au cours du développement embryonnaire. *C. Séanc. Soc. Biol.*, **142**, 1241–54.

34. BRACHET, J. (1950). *Chemical Embryology.* Interscience Publ., New York and London.

35. BRACHET, J. (1957). *Biochemical Cytology.* Academic Press, New York.

36. BRACHET, J. (1959). Tissue interactions: embryonic induction. Discussion. In *Biological organisation, cellular and subcellular*, ed. WADDINGTON, C. H., pp. 225–30. Pergamon Press, London.

37. BRACHET, J. (1960). *The Biochemistry of Development.* Pergamon Press, London.

38. BRACHET, J. (1960). Nucleic acids and growth. In *Fundamental Aspects of Normal and Malignant Growth*, ed. NOWINSKI, W., pp. 260–304. Elsevier, Amsterdam.

39. BRACHET, J. (1962). *Nucleic acids in development. J. cell. comp. Physiol.*, Suppl. 1, **60**, 1.

40. BRACHET, J. (1965). The role of nucleic acids in morphogenesis. *Prog. Biophys. molec. Biol.*, **15**, 99–127.

41. BRACHET, J. (1968). Synthesis of macromolecules and morphogenesis in *Acetabularia*. In *Current Topics in Developmental Biology*, eds. MOSCONA, A. A. and MONROY, A., Vol. 3, pp. 1–35. Academic Press, New York.

42. BRACHET, J. and DENIS, H. (1963). Effects of actinomycin D on morphogenesis. *Nature, Lond.*, **198**, 205–206.

43. BRACHET, J., DENIS, H. and DE VITRY, F. (1964). The effects of actinomycin D and puromycin on morphogenesis in amphibian eggs and *Acetabularia mediterranea*. *Devl Biol.*, **9**, 398–434.

44. BRADEN, A. W. H., AUSTIN, C. R. and DAVID, H. A. (1954). The reaction of the zona pellucida to sperm penetration. *Aust. J. biol. Sci.*, **7**, 391–409.

45. BRIGGS, R. and JUSTUS, J. T. (1968). Partial characterization of a component from normal eggs which corrects the maternal effect of gene O in the Mexican Axolotl (*Amblystoma mexicanum*). *J. exp. Zool.*, **167**, 105–15.

46. BRIGGS, R. and KING, T. (1952). Transplantation of living nuclei from blastula cells into enucleated frog eggs. *Proc. natn. Acad. Sci. U.S.A.*, **38**, 455–63.

47. BRIGGS, R. and KING, T. J. (1953). Factors affecting the transplantability of nuclei of frog embryonic cells. *J. exp. Zool.*, **122**, 485–506.

48. BRIGGS, R. and KING, T. J. (1957). Changes in the nuclei of differentiating endoderm cells as revealed by nuclear transplantation. *J. Morph.*, **100**, 269–312.

49. BRIGGS, R. and KING, T. J. (1959). Nucleocytoplasmic interactions in eggs and embryos. In *The Cell*, eds. BRACHET, J. and MIRSKY, E., Vol. 1, pp. 537–617. Academic Press, New York.

50. BRIGGS, R. and KING, T. J. (1960). Nuclear transplantation studies on the

early gastrula (*Rana pipiens*). I. Nuclei of presumptive endoderm. *Devl Biol.*, **2**, 252–70.

51. BRIGGS, R., KING, T. J. and DI BERARDINO, M. A. (1961). Development of nuclear-transplant embryos of known chromosome complement following parabiosis with normal embryos. In *Symposium on Germ Cells and Development*, ed. RANZI, S., pp. 441–77. Inst. Intern. d'Embryologie and Fondazione A. Baselli, Milan, Italy.

52. BROWN, D. D. (1964). RNA sythesis during amphibian development. *J. Exp. Zool.*, **157**, 101–14.

53. BROWN, D. D. (1967). The genes for ribosomal RNA and their transcription during amphibian development. In *Current Topics in Developmental Biology*, eds. MOSCONA, A. A. and MONROY, A. Vol. 2, pp. 47–73. Academic Press, New York.

54. BROWN, D. D. and CASTON, J. D. (1962). Biochemistry of amphibian development. I. Ribosome and protein synthesis in early development of *Rana pipiens*. *Devl Biol.*, **5**, 412–34.

55. BROWN, D. D. and GURDON, J. B. (1964). Absence of ribosomal RNA synthesis in the anucleolate mutant of *Xenopus laevis*. *Proc. natn Acad. Sci. U.S.A.*, **51**, 139–46.

56. BROWN, D. D. and LITTNA, E. (1964). RNA synthesis during the development of *Xenopus laevis*, the South African 'clawed toad'. *J. molec. Biol.*, **8**, 669–87.

57. BROWN, D. D. and LITTNA, E. (1964). Variations in the synthesis of stable RNA's during oogenesis and development of *Xenopus laevis*. *J. molec. Biol.*, **8**, 688–95.

58. CARTER, T. C. (1954). Genetics of luxate mice. IV. Embryology. *J. Genet.*, **52**, 1–35.

59. CHURCH, R. B. (1970). Differential gene activity. In *Congenital Malformations*, eds. FRASER, F. C. and MCKUSICK, V. A., pp. 19–28. Proc. 3rd Int. Conf. on Cong. Malformations, The Hague, Netherlands. *Excerpta med*, (*Amst.*)

60. CLEVER, U. (1966). Gene activity patterns and cellular differentiation. *Am. Zool.*, **6**, 33–41.

61. COLLIER, J. R. (1966). The transcription of genetic information in the spiralian embryo. In *Current Topics in Developmental Biology*, eds. MOSCONA, A. A. and MONROY, A., Vol. 1, pp. 39–58. Academic Press, New York.

62. COLWIN, A. L. and COLWIN, L. H. (1957). *Morphology of fertilization*: acrosome filament formation and sperm entry. In *The Beginnings of Embryonic Development*, eds. TYLER, A., VON BORSTEL, R. C. and METZ, C. B. pp. 135–68. Am. Ass. Advance. Sci., Washington, D.C.

63. COLWIN, A. L. and COLWIN, L. H. (1960). Fine structure studies of fertilization with special reference to the role of the acrosomal region of the spermatozoon during penetration of the egg (*Hydroides hexagonus Annelida*). In *Symposium on the Germ Cells and Earliest Stages of Development*, pp. 220–22. Institut Intern. d'Embryologie and Fondazione A. Baselli, Instituo Lombardo, Milan, Italy.

64. COLWIN, A. L. and COLWIN, L. H. (1963). Role of the gamete membranes in fertilization in *Saccoglossus kowalevskii* (*Enteropneusta*). I. The acrosomal region and its changes in early stages of fertilization. *J. Cell Biol.*, **19**, 477–500.
65. COLWIN, A. L. and COLWIN, L. H. (1964). Role of the gamete membranes in fertilization. In *Cellular Membranes in Development*, ed. LOCKE, M., pp. 233–79. Academic Press, New York.
66. COLWIN, L. H. and COLWIN, A. L. (1963). Role of the gamete membranes in fertilization in *Saccoglossus kowalevskii* (*Enteropneusta*). II. Zygote formation by membrane fusion. *J. Cell Biol.*, **19**, 501–18.
67. COLWIN, L. H. and COLWIN, A. L. (1967). Membrane fusion in relation to sperm-egg association. In *Fertilization*, eds., METZ, C. B. and MONROY, A., Vol. 1, pp. 295–367. Academic Press, New York.
68. COULOMBRE, J. L. and COULOMBRE, A. J. (1963). Lens development: fiber elongation and lens orientation. *Science, N.Y.*, **142**, 1489–90.
69. DAGG, C. P. (1967). Combined action of fluorouracil and two mutant genes on limb development in the mouse. *J. exp. Zool.* **164**, 479–89.
70. DAN, J. (1954). Studies on the acrosome. II. Acrosome reaction in starfish spermatozoa. *Biol. Bull. mar. biol. Lab.*, *Woods Hole*, **107**, 203–18.
71. DAN, J. C. (1956). The acrosome reaction. *Int. Rev. Cytol.*, **5**, 365–93.
72. DAN, J. C. (1960). Studies on the acrosome. VI. Fine structure of the starfish acrosome. *Expl Cell Res.*, **19**, 13–28.
73. DAN, J. C. (1967). Acrosome reaction and lysins. In *Fertilization I*, eds. METZ, C. B. and MONROY, A., pp. 237–80. Academic Press, New York.
74. DAVIDSON, E. (1964). Gene activity in differentiated cells. In *Retention of Functional Differentiation in Cultured Cells*. Wistar Symp. Monogr., **1**, 49–59.
75. DAVIDSON, E. H. (1968). *Gene Activity in Early Development*. Academic Press, New York.
76. DE HAAN, R. L. and URSPRUNG, H. (1965). *Organogenesis*. Holt, Rinehart and Winston, New York.
77. DENIS, H. (1964). Effects de l'actinomycine sur le développement embryonnaire. I. Etude morphologique: suppression par l'actinomycine de la compétence de l'ectoderme et du pouvoir inducteur de la lèvre blastoporale. *Devl. Biol.*, **9**, 435–57.
78. DENIS, H. (1968). The role of messenger ribonucleic acid in embryonic development. In *Advances in Morphogenesis 7*, eds. ABERCROMBIE, J., BRACHET, J., and KING, T., pp. 115–148. Academic Press, New York.
79. DEUCHAR, E. M. (1966). *Biochemical Aspects of Amphibian Development*. Methuen and Company, Ltd, London.
80. DI BERARDINO, M. A. and KING, T. J. (1967). Development and cellular differentiation of neural nuclear transplants of known karyotype. *Devl. Biol.*, **15**, 102–28.
81. DRIESCH, H. (1892). The potency of the first two cleavage cells in echinoderm development. Experimental production of partial and double formations. In *Foundations of Experimental Embryology*, eds.,

WILLIER, B. H. and OPPENHEIMER, J. M., pp. 38–50. Prentice-Hall, Englewood Cliffs, N.J.

82. DRIESCH, H. (1897). Betrachtungen uber die Organisation des Eies und ihre Genese. *Arch. EntwMech. Org.*, **4**, 75–124.

83. DUNN, L. C. (1939/1940). Heredity and development of early abnormalities in vertebrates. *Harvey Lectures*, **35**, 115–65.

84. EBERT, J. D. (1965). *Interacting Systems in Development*, Modern Biology Series. Holt, Rinehart and Winston, New York.

85. ELSDALE, T. R. FISCHBERG, M. and SMITH, S. (1958). A mutation that reduces nucleolar number in *Xenopus laevis*. *Expl. Cell Res.*, **14**, 642–43.

86. FICQ. A. (1954). Analyse de l'induction neurale par autoradiographie. *Experientia*, **10**, 20–21.

87. FICQ, A. (1954). Analyse de l'induction neurale chez les Amphibiens au moyen de d'organisateurs marqués. *J. Embryol. exp. Morph.* **2**, 194–203.

88. FISCHBERG, M. and BLACKLER, A. W. (1961). How cells specialize. *Scient. Am.*, **205**, 124–40 (#3).

89. FISCHBERG, M., GURDON, J. B. and ELSDALE, T. R. (1958). Nuclear transfer in Amphibia and the problem of the potentialities of the nuclei of differentiating tissues. *Expl Cell Res.* Suppl., **6**, 161–78.

90. FLICKINGER, R. A., GREENE, R., KOHL, D. M. and MIYAGI, M. (1966). Patterns of synthesis of DNA-like RNA in parts of developing frog embryos. *Proc. natn. Acad. Sci.*, *U.S.A.*, **56**, 1712–18.

91. GALL, J. G. (1963). Chromosomes and cytodifferentiation. In *Cytodifferentiation and Macromolecular Synthesis*, ed. LOCKE, M., pp. 119–43. Academic Press, New York.

92. GALL, J. G. and CALLAN, H. G. (1962). H^3 uridine incorporation of lampbrush chromosomes. *Proc. natn. Acad. Sci. U.S.A.* **48**, 562–70.

93. GLUECKSOHN-SCHOENHEIMER, S. (1938). The development of two tailless mutants in the house mouse. *Genetics, Princeton*, **23**, 573–84.

94. GLUECKSOHN-SCHOENHEIMER, S. (1945). The embryonic development of mutants of the Sd strain in mice. *Genetics, Princeton*, **30**, 29–38.

95. GLUECKSOHN-WAELSCH, S. (1954). Some genetic aspects of development. *Cold Spring Harb. Symp. quant. Biol.*, **19**, 41–49.

96. GOETINCK, P. F. (1966). Genetic aspects of skin and limb development. In *Current Topics in Developmental Biology*, eds., MOSCONA, A. A. and MONROY, A., pp. 253–83. Academic Press, New York.

97. GOLDSCHMIDT, R. (1938). *Physiological Genetics*. McGraw-Hill.

98. GOSS, R. J. (1963). *Adaptive Growth*. Logos Press, London.

99. GRANT, P. (1965). Informational molecules and embryonic development. In *The Biochemistry of Animal Development*, ed. WEBER, R., Vol. 1, pp. 483–583. Academic Press, New York.

100. GRAY, G. W. (1957). The Organizer. *Scient. Am.*, **197**, 79–88.

101. GROBSTEIN, C. (1953). Morphogenic interaction between embryonic mouse tissue separated by a membrane filter. *Nature, Lond.*, **172**, 869–71.

102. GROBSTEIN, C. (1953). Epitheliomesenchymal specificity in the mor-

phogenesis of mouse submandibular rudiments *in vitro. J. exp. Zool.*, **124**, 383–414.

103. GROBSTEIN, C. (1954). Tissue interaction in the morphogenesis of mouse embryonic rudiments *in vitro*. In *Aspects of Synthesis and Order in Growth*, ed. RUDNICK, D., pp. 233–56. *13th Symp. Devl. Growth*, Princeton University Press, Princeton, New Jersey.

104. GROBSTEIN, C. (1955). Inductive interaction in the development of the mouse metanephros. *J. exp. Zool.*, **130**, 319–40.

105. GROBSTEIN, C. (1956). Transfilter induction of tubules in mouse metanephrogenic mesenchyme. *Expl Cell Res.*, **10**, 424–40.

106. GROBSTEIN, C. (1956). Inductive tissue interaction in development. *Adv. Cancer Res.*, **4**, 187–236.

107. GROBSTEIN, C. (1959). Differentiation of vertebrate cells. In *The Cell*, eds. BRACHET, J. and MIRSKY, A., Vol. 1, pp. 437–96. Academic Press, New York.

108. GROBSTEIN, C. (1959). Autoradiography of the interzone between tissues in inductive interaction. *J. exp. Zool.*, **142**, 203–14.

109. GROBSTEIN, C. (1960). Passage of radioactivity into a membrane filter from spinal cord pre-incubated with tritiated amino acids or nucleosides. In *La Culture Organotypique*, associations et dissociations d'organes en culture *in vitro*, ed. WOLFF, M. E., No. 101, pp. 169–82. Editions du Centre National de la Recherche Scientifique, Paris.

110. GROBSTEIN, C. (1961). Cell contact in relation to embryonic induction. *Expl Cell Res.* Suppl., **8**, 234–45.

111. GROBSTEIN, C. (1964). Cytodifferentiation and its controls. *Science, N.Y.*, **143**, 643–50.

112. GROBSTEIN, C. and COHEN, J. (1965). Collagenase. Effect on the morphogenesis of embryonic salivary epithelium *in vitro. Science, N.Y.*, **150**, 626–28.

113. GROBSTEIN, C. and DALTON, A. J. (1957). Kidney tubule induction in mouse metanephrogenic mesenchyme without cytoplasmic contact. *J. exp. Zool.*, **135**, 57–73.

114. GROSS, P. R. (1964). The immediacy of genomic control during early development. *J. expl. Zool.*, **157**, 21–38.

115. GROSS, P. R. (1967). The control of protein synthesis in embryonic development and differentiation. In *Current Topics of Developmental Biology*, eds. MOSCONA, A. A. and MONROY, A., Vol. 2, pp. 1–46. Academic Press, New York.

116. GROSS, P. R. and COUSINEAU, G. H. (1963). Effects of actinomycin D on macromolecule synthesis and early development in sea urchin eggs. *Biochem. biophys. Res. Commun.*, **10**, 321–26.

117. GROSS, P. R. and COUSINEAU, G. H. (1964). Macromolecule synthesis and the effect of actinomycin on early development. *Expl Cell Res.*, **33**, 368–95.

118. GROSS, P. R., MALKIN, L. I. and MOYER, W. A. (1964). Templates for the first proteins of embryonic development. *Proc. natn. Acad. Sci. U.S.A.*, **51**, 407–14.

119. GRÜNEBERG, H. (1952). *The Genetics of the Mouse*. Martinus Nijhoff, The Hague, Netherlands.

120. GRÜNEBERG, H. (1963). *The Pathology of Development I*. J. Wiley, New York.

121. GURDON, J. B. (1960). The developmental capacity of nuclei taken from differentiating endoderm cells of *Xenopus laevis*. *J. Embryol. exp. Morph.*, **8**, 505–26.

122. GURDON, J. B. (1962). Adult frogs derived from the nuclei of single somatic cells. *Devl Biol.*, **4**, 256–73.

123. GURDON, J. B. (1969). Intracellular communication in early animal development. In *Communication in Development*. 28th Symp. of the Soc. for Devl. Biol., pp. 59–82. Academic Press, New York.

124. GURDON, J. B. (1964). The transplantation of living cell nuclei. In *Advances in Morphogenesis*, eds. ABERCROMBIE, M. and BRACHET, J., Vol. 5, pp. 1–43. Academic Press, New York.

125. GURDON, J. B. (1967). Nuclear transplantation and cell differentiation. In *Cell Differentiation, CIBA Fdn Symp.*, eds. DE REUCK, A.V.S. and KNIGHT, Little, Brown and Company, Boston.

126. GURDON, J. B. and BROWN, D. (1965). Cytoplasmic regulations of RNA synthesis in nucleolus formation in developing embryos of *Xenopus laevis*. *J. molec. Biol.*, **12**, 27–35.

127. HADEK, R. (1963). Submicroscopic changes in the penetrating spermatozoon of the rabbit. *J. Ultrastruct. Res.*, **8**, 161–69.

128. HADORN, E. (1948). Gene action in growth and differentiation of lethal mutants of *Drosophila*. *Symp. Soc. exp. Biol.* **2**, 177–95.

129. HADORN, E. (1961). *Developmental Genetics and Lethal Factors*. Methuen, London.

130. HAMBURGH, M. (1954). Embryology of trypan blue induced abnormalities in mice. *Anat. Rec.*, **119**, 409–27.

131. HAMBURGH, M. (1960). Observations on the neuropathology of 'Reeler', a neurological mutation in the mouse. *Experientia*, **16**, 460–61.

132. HAMBURGH, M. (1963). Analysis of the postnatal developmental effects of 'Reeler', a neurological mutation in mice. A study in developmental genetics. *Devl Biol.*, **8**, 165–85.

133. HAMBURGH, M. (1969). Thyroid and growth hormone in neurogenesis. In *Current Topics of Developmental Biology*, Vol. 4, eds., MONROY, A. and MOSCONA, A. A., pp. 109–49. Academic Press, New York.

134. HAMBURGH, M., BURKART, J. and WEIL, F. (1971). Thyroid sensitive targets of the developing nervous system. In *Proceedings of an International Symposium on Hormones in Development*, eds. HAMBURGH, M. and BARRINGTON, E. J. W. Appleton-Century-Crofts, New York.

135. HAMBURGH, M. and CALLAHAN, V. (1967). Differences in teratogenic response in capacity to repair in embryos of two inbred strains. *Experientia*, **23**, 1–4.

136. HÄMMERLING, J. (1966). Nucleocytoplasmic relationships in the development of *Acetabularia*. In *Developmental Biology*, ed. FLICKINGER, R., pp. 23–47. W. C. Brown, Dubuque, Iowa.

137. HARRIS, H. (1968). *Nucleus and Cytoplasm*. Clarendon Press, Oxford.
138. HARRISON, R. G. (1918). Experiments on the development of the fore-limb of Amblystoma, a self-differentiating equipotential system. *J. exp. Zool.*, **25**, 413–61.
139. HAYASHI, Y. (1956). Morphogenetic effects of pentose nucleoprotein from the liver upon the isolated ectoderm. *Embryologia*, **3**, 57–67.
140. HAYASHI, Y. (1958). The effects of pepsin and trypsin on the inductive ability of pentose nucleoprotein from guinea pig liver. *Embryologia*, **4**, 33–53.
141. HAYASHI, Y. (1959). The effect of ribonuclease on the inductive ability of liver pentose nucleoprotein. *Devl Biol.*, **1**, 247–68.
142. HIRAMOTO, Y., (1962). An analysis of the mechanism of fertilization by means of enucleation of sea urchin eggs. *Expl Cell Res.*, **28**, 323–34.
143. HOLTFRETER, J. (1933). Nachweis der Induktionsfähigkeit abgetöteter Keimteile. Isolations und Transplantationsversuche. *Arch. EntwMech. Org.*, **128**, 584–633.
144. HOLTFRETER, J. (1944). Neural differentiation of ectoderm through exposure to saline solution. *J. exp. Zool.*, **95**, 307–40.
145. HOLTFRETER, J. (1947). Neural induction in explants which have passed through a sublethal cytolysis. *J. exp. Zool.*, **106**, 197–222.
146. HOLTFRETER, J. (1948). Concepts on the mechanism of embryonic induction and its relation to parthenogenesis and malignancy. *Symp. Soc. exp. Biol.*, **2**, 17–48.
147. HOLTFRETER, J. (1951). Some aspects of embryonic induction. *Growth* (suppl), **10**, 117–52.
148. HOLTFRETER, J. (1955). Studies on the diffusability, toxicity and patho-genic properties of 'inductive' agents derived from dead tissues. *Expl Cell Res.*, (suppl.), **3**, 188–209.
149. HOLTFRETER, J. (1968). Mesenchyme and epithelia in inductive and morphogenetic processes. In *Epithelial–Mesenchymal Interactions*, ed. FLEISCHMAJER, R. and BILLINGHAM, R. pp. 1–29. The Williams and Wilkins Company, Baltimore.
150. HOLTFRETER, J. and HAMBURGER, V. (1955). Embryogenesis: Progressive Differentiation. Amphibians. In *Analysis of Development*, eds. WILLIER, B. H., WEISS, P. A. and HAMBURGER, V., pp. 230–96. Saunders, Phila-delphia and London.
151. HOLTZER, H. (1961). Aspects of chondrogenesis and myogenesis. In *Molecular and Cellular Synthesis*, 19th Growth Symp., ed. RUDNIK, D. Ronald Press, New York.
152. HOLTZER, H. (1963). Induktion und Morphogenese, 13th Mosbacher *Colloquium Ges. physiol. Chem.* Julius Springer, Heidelberg.
153. HOLTZER, H. (1964). Control of chondrogenesis in the embryo. *Biophys. J.*, **4**, 239–50.
154. HOLTZER, H. (1968). Induction of chondrogenesis. A concept in quest of mechanisms. In *Epithelial–Mesenchymal Interactions*, eds. FLEISCH-MAJER, R. and BILLINGHAM, R., pp. 152–62. The Williams and Wilkins Company, Baltimore.

155. HOLTZER, H. and DETWILER, S. (1953). III. Induction of skeletogenous cells. *J. exp. Zool.*, **123**, 335–69.
156. HORWITZ, B. A. (1965). Rates of oxygen consumption of fertilized and unfertilized *Asterias, Arbacia* and *Spisula* eggs. *Expl Cell Res.*, **38**, 620–25.
157. HULTIN, T. (1950). The protein metabolism of sea urchin eggs during early development studied by means of N-labeled ammonia. *Expl Cell Res.*, **1**, 599–602.
158. HULTIN, T. (1952). Incorporation of N-labeled glycine and alanine into the proteins of developing sea urchin eggs. *Expl Cell Res.*, **3**, 494.
159. HULTIN, T. (1961). Activation of ribosomes in sea urchin eggs in response to fertilization. *Expl Cell Res.*, **25**, 405.
160. HUMPHREY, R. R. (1966). A recessive factor (*o*, for ova deficient) determining a complex of abnormalities in the Mexican axolotl (*Amblystoma mexicanum*). *Devl Biol.*, **13**, 57–76.
161. JACOBSON, A. G. (1966). Inductive processes in embryonic development. *Science, N.J.*, **152**, 25–34.
162. JOHNEN, A. G. (1956). Experimental studies about the relationships in the induction process. I. Experiments on *Amblystoma mexicanun*. *Proc. Acad. Sci., Amsterdam*, Series C, **59**, 554–61.
163. KALLMAN, F. and GROBSTEIN, C. (1964). Fine structure of differentiating mouse pancreatic exocrine cells in transfilter culture. *J. Cell Biol.*, **20**, 399–413.
164. KALLMAN, F. and GROBSTEIN, C. (1965). Source of collagen at epitheliomesenchymal interfaces during inductive interaction. *Devl Biol.*, **11**, 169–83.
165. KALLMAN, F. and GROBSTEIN, C. (1966). Localization of glucosamine-incorporating materials at epithelial surfaces during salivary epitheliomesenchymal interaction *in vitro*. *Devl Biol.*, **14**, 52–67.
166. KERR, N. S. (1968). *Principles of Development*, Concepts of Biology Series. William C. Brown Co., Dubuque, Iowa.
167. KING, T. J. (1955). Changes in the nuclei of differentiating gastrula cells, as demonstrated by nuclear transplantation. *Proc. natn. Acad. Sci., U.S.A.*, **41**, 321–25.
168. KING, T. J. and BRIGGS, R. (1956). Serial transplantation of embryonic nuclei. *Cold Spring Harb. Symp. quant. Biology*, **21**, 217–90.
169. KOCH, W. E. and GROBSTEIN, C. (1963). Transmission of radioisotopically labelled materials during embryonic induction *in vitro*. *Devl Biol.*, **7**, 303–23.
170. LANDAUER, W. (1932). Studies on the Creeper fowl. III. The early development and lethal expression of homozygous Creeper embryos. *J. Genet.*, **25**, 367–94.
171. LANDAUER, W. (1948). Hereditary abnormalities and their chemically induced phenocopies. *Growth* Symp., **12**, 171–200.
172. LANDAUER, W. (1954). On the chemical production of developmental abnormalities and of phenocopies in chicken embryos. *J. cell. comp. Physiol.*, suppl. 1, **43**, 261–305.

173. LANDAUER, W. (1959). The phenocopy concept: illusion or reality. *Experientia*, **15**, 409–12.

174. LASH, J. W. (1963). Tissue interaction and specific metabolic responses. Chondrogenic induction and differentiation. In *Cytodifferentiation and Macromolecular Synthesis*, 21st Symp. Dev. Growth, ed. LOCKE, M. pp. 235–60. Academic Press, New York.

175. LASH, J. W. (1967). Differential behaviour of anterior and posterior embryonic chick somites *in vitro*. *J. exp. Zool.*, **165**, 47–56.

176. LASH, J. W. (1968). Somitic mesenchyme and its response to cartilage induction. In *Epithelial–Mesenchymal Interactions*, eds. FLEISCHMAJER, R. and BILLINGHAM, R., pp. 165–71. The Williams and Wilkins Company, Baltimore.

177. LASH, J. W. (1968). Chondrogenesis, genotypic and phenotypic expression. In *Symposium on Molecular Aspects of Differentiation*. *J. Cell. Physiol.*, Suppl. 1, **72**, 45–46.

178. LASH, J. W., GLICK, M. C. and MADDEN, J. W. (1964). Cartilage induction *in vitro* and sulfate-activating enzymes. In *Metabolic Control in Animal Cells*. Natn. Cancer Inst. Monogr. No. **13**, 39–49.

179. LASH, J. W., HOLTZER, S. and HOLTZER, H. (1957). An experimental analysis of the development of the spinal column. VI. Aspects of cartilage induction. *Expl Cell Res.*, **13**, 292–303.

180. LASH, J., HOLTZER, H. and WHITEHOUSE, M. (1960). The uptake of radioactive sulfate during cartilage induction. *Devl Biol.*, **2**, 76–89.

181. LASH, J. W., HOMMES, F. A. and ZILLIKEN, F. (1962). Induction of cell differentiation. I. The *in vitro* induction of vertebral cartilage with a low-molecular-weight tissue component. *Biochim. biophys. Acta*, **56**, 313–19.

182. LEHMANN, F. E. (1938). Regionale Verschiedenheiten des Organisators von Triton, insbesondere in der vorderen und hinteren Kopf-region nachgewiesen durch phasenspezifische Erzengung von Lithiumbedingten und operativ bewirkten Regional-defekten. *Arch. EntwMech. Org.*, **138**, 106–58.

183. LEVI-MONTALCINI, R. and ANGELETTI, P. U. (1965). The action of nerve growth factor on sensory and sympathetic cells. In *Organogenesis*, eds. DEHAAN, R. L. and URSPRUNG, H., pp. 187–98. Holt, Rinehart and Winston, New York.

184. LILLIE, F. R. (1912). The production of sperm iso-agglutinins by ova. *Science, N.Y.*, **36**, 527–30.

185. LILLIE, F. R. (1913). The mechanism of fertilization. *Science, N.Y.*, **38**, 524–29.

186. LILLIE, F. R. (1914). Studies of fertilization. VI. The mechanism of fertilization in *Arbacia*, *J. exp. Zool.*, **16**, 523–90.

187. LILLIE, F. R. (1919). *Problems of Fertilization*. University of Chicago Press, Chicago, Illinois.

188. LUNDBLAD, G. (1949). Proteolytic activity in eggs and sperms from sea urchins. *Nature, Lond.*, **163**, 643.

189. LUNDBLAD, G. (1950). Proteolytic activity in sea urchin gametes. *Expl Cell Res.*, **1**, 264–71.

190. MARZULLO, G. and LASH, J. W. (1967). Separation of phosphorylated and UDP derivatives of hexosamines and acetylhexosamines by TLC. *Analyt. Biochem.*, **18**, 579–82.

191. MARZULLO, G. and LASH, J. W. (1967). Acquisition of chondrocyte phenotype. *Exp. Biol. and Med.*, **1**, 213–18.

192. METZ, C. B. (1957). Mechanisms in fertilization. In *Physiological Triggers*, ed. BULLOCK, T. H., pp. 17–45. American Physiological Society, Washington, D.C.

193. METZ, C. B. (1957). Specific egg and sperm substances and activation of the egg. In *Beginnings of Embryonic Development*, eds. TYLER, A., VON BORSTEL, R. C. and METZ, C. B., pp. 23–69. Am. Ass. Advance. Sci., Washington, D.C.

194. METZ, C. B. (1967). Gamete surface components and their role in fertilization. In *Fertilization, Comparative Morphology, Biochemistry and Immunology I*, eds. METZ, C. B. and MONROY, A., pp. 163–224. Academic Press, New York.

195. MINTZ, B. (1964). Synthetic processes and early development in the mammalian egg. *J. exp. Zool.*, **157**, 85–100.

196. MIRSKY, A. E. (1964). Regulation of genetic expression, *J. exp. Zool.*, **157**, 45–48.

197. MONROY, A. (1965). Biochemical aspects of fertilization. In *The Biochemistry of Animal Development*, ed. WEBER, R., pp. 73–128. Academic Press, New York.

198. MONROY, A. (1965). *Chemistry and Physiology of Fertilization.* Holt, Rinehart and Winston, New York.

199. MONROY, A. and GROSS, P. R. (1967). The control of gene action during echinoderm embryogenesis. *Exp. Biol. Med.*, **7**, 37–51.

200. MONROY, A. and TYLER, A. (1967). The activation of the egg. In *Fertilization I*, eds. METZ, C. B. and MONROY, A., pp. 369–403. Academic Press, New York.

201. MOORE, J. (1960). Serial backtransfers of nuclei in experiments involving two species of frogs. *Devl Biol.*, **2**, 535–50.

202. MOTOMURA, I. (1953). Secretion of sperm agglutinin in the fertilized eggs of sea urchins. *Expl Cell Res.*, **5**, 187–90.

203. NAKANO, E., GIUDICE, G., and MONROY, A. (1958). On the incorporation of S^{35}-methionine in artificially activated sea urchin eggs. *Experientia*, **14**, 11–13.

204. NAKANO, E. and MONROY A. (1958). Incorporation of S^{35}-methionine in the cell fractions of sea urchin eggs and embryos. *Expl Cell Res.*, **14**, 236–44.

205. NEEDHAM, J. (1931). *Chemical Embryology I–III.* University Press, Cambridge.

206. NEEDHAM, J. (1950). *Biochemistry and Morphogenesis.* University Press, Cambridge.

207. NEEDHAM, J., WADDINGTON, C. H. and NEEDHAM, D. M. (1934). Physico-chemical experiments on the amphibian organizer. *Proc. R. Soc.* Series B, **114**, 393–422.

208. NEMER, M. (1963). Old and new RNA in the embryogenesis of the purple sea urchin. *Proc. natn. Acad. Sci. U.S.A.*, **50**, 230–35.

209. NEMER, M. and BARD, S. G. (1963). Polypeptide synthesis in sea urchin embryogenesis: an examination with synthetic polyribonucleotides. *Science, N.Y.*, **140**, 664–66.

210. NIEUWKOOP, P. D. (1966). Induction and pattern formation as primary mechanisms in early embryonic differentiation. In *Cell Differentiation and Morphogenesis*, Int. Lecture Course, Wageningen, The Netherlands, pp. 120–35. North Holland Publishing Co., Amsterdam.

211. NIU, M. C. (1956). New approaches to the problem of induction. In *Cellular Mechanisms in Differentiation and Growth*, ed. RUDNICK, D., pp. 155–71. University Press, Princeton, New Jersey.

212. NIU, M. C. (1958). The role of ribonucleic acid in embryonic differentiation. *Anat. Rec.*, **131**, 585.

213. NIU, M. C. (1958). Thymus ribonucleic acid and embryonic differentiation. *Proc. natn. Acad. Sci. U.S.A.*, **44**, 1264–74.

214. NIU, M. C. and TWITTY, V. C. (1953). The differentiation of gastrula ectoderm in medium conditioned by axial mesoderm. *Proc. natn. Acad. Sci. U.S.A.*, **39**, 985–89.

215. OHNISHI, T. and SUGIYAMA, M. (1963). Polarographic studies of oxygen uptake of sea urchin eggs. *Embryologia*, **8**, 79–88.

216. PATTEN, B. M. (1964). *Foundations of Embryology*, 2nd ed. McGraw-Hill, New York.

217. POULSON, D. F. (1940). The effects of certain X-chromosome deficiencies on the embryonic development of *Drosophila melanogaster*. *J. exp. Zool.*, **83**, 271–326.

218. RAVEN, C. P. (1959). *An Outline of Developmental Physiology*. Pergamon Press, London.

219. REYER, R. W. (1962). Regeneration in the amphibian eye. In *Regeneration*, 20th Symp. Dev. Growth, ed. RUDNICK, D., pp. 211–65. The Ronald Press Co., New York.

220. ROTHSCHILD, LORD (1949). The metabolism of fertilized and unfertilized sea urchin eggs. The action of light and carbon monoxide. *J. exp. Biol.*, **26**, 100–11.

221. ROTHSCHILD, LORD (1956). *Fertilization*. Methuen, London.

222. ROTHSCHILD, LORD and SWANN, M. M. (1951). The conduction time of the block to polyspermy in the sea urchin egg. *Expl Cell Res.*, **2**, 137.

223. ROUNDS, D. E. and FLICKINGER, R. E. (1958). Distribution of ribonucleoprotein during neural induction of the frog embryo. *J. exp. Zool.*, **137**, 479–500.

224. ROUX, W. (1888). Contributions to the developmental mechanics of the embryo. On the artificial production of half embryos by destruction of one of the first two blastomeres, and the later development (post-generation) of the missing half of the body. In *Foundations of Experimental Embryology*, eds. WILLIER, B. H. and OPPENHEIMER, J. M., pp. 3–37. Prentice-Hall, Englewood Cliffs, N.J.

225. RUDNIK, D. (1945). Differentiation of prospective limb material from Creeper chick embryos in coelomic grafts. *J. exp. Zool.*, **100**, 1–17.

226. RUNNSTROM, J. (1949). The mechanism of fertilization in metazoa. In *Advances in Enzymology*, Vol. 9, pp. 241–327. Wiley (Interscience), New York.

227. RUNNSTROM, J., HAGSTROM, B. E. and PERLMANN, P. (1959). Fertilization. In *The Cell*, eds. BRACHET, J. and MIRSKY, A. E., Vol. 1, pp. 327–97. Academic Press, New York.

228. RUSSELL, E. S. (1949). Analysis of pleiotropism at the W-locus in the mouse: Relationship between the effects of W and W^v substitution on hair pigmentation and on erythrocytes. *Genetics*, **34**, 708–24.

229. RUTTER, W. J., KEMP, J. D., BRADSHAW, W. S., CLARK, W. R., RONZIO, R. A. and SANDERS, J. T. (1968). Regulation of specific protein synthesis in cytodifferentiation. *J. Cell Physiol.*, Suppl. 1, **72**, 1–18.

230. RUTTER, W. J., WESSELLS, N. K. and GROBSTEIN, C. (1964). Control of specific synthesis in the developing pancreas. *Natn. Cancer Inst. Monogr.* **13**, 51–64.

231. SAUNDERS, J. W., JR. (1948). The proximo-distal sequence of origin of the parts of the chick wing and the role of the ectoderm. *J. exp. Zool.*, **108**, 363–403.

232. SAUNDERS, J. W., JR. (1968). *Animal Morphogenesis*, Current Concepts in Biology Series. The Macmillan Company, New York.

233. SAUNDERS, J. W., JR. and GASSELING, M. T. (1963). Trans-filter propagation of apical ectoderm maintenance factor in the chick embryo wing bud. *Devl Biol.*, **7**, 64–78.

234. SAUNDERS, J. W., JR., GASSELING, M. T. and GFELLER, M. D. SR. (1958). Interactions of ectoderm and mesoderm in the origin of axial relationships in the wing of the fowl. *J. exp. Zool.*, **137**, 39–74.

235. SAXEN, L. (1961). Transfilter neural induction of amphibian ectoderm. *Devl Biol.*, **3**, 140–52.

236. SAXEN, L. and TOIVONEN, S. (1962). *Primary Embryonic Induction*. Logos Press, London.

237. SIRLIN, T. L., BRAHMA, S. K. and WADDINGTON, C. H. (1956). Studies on embryonic induction using radioactive traces. *J. Embryol. exp. Morph.*, **4**, 248–53.

238. SISKEN, B. F. and GLUECKSOHN-WAELSCH, S. (1959). A developmental study of the mutation phocomelia in the mouse. *J. exp. Zool.*, **142**, 623–42.

239. SMITH, L. J. (1956). A morphological and histochemical investigation of a preimplantation lethal (t^{12}) in the house mouse. *J. exp. Zool.*, **132**, 51–83.

240. SMITH, L. J. and STEIN, K. F. (1962). Axial elongation in the mouse and its retardation in homozygous looptail mice. *J. Embryol. exp. Morph.*, **10**, 73–87.

241. SPEMANN, H. (1918). Uber die Determination der ersten Organanlagen des Amphibienembryo. I-VI. *Arch. EntwMech. Org.*, **43**, 448–555.

242. SPEMANN, H. (1938). *Embryonic Development and Induction*. Yale University Press, New Haven, Connecticut.

243. SPEMANN, H. and MANGOLD, H. (1924). Uber Induktion von Embryo-nalanlagen durch Implantation artfremder Organisatoren. *Arch. EntwMech. Org.*, **100**, 599–638.

244. SPEMANN, H. and MANGOLD, H. (1924). Induction of embryonic primordia by implantation of organizers from a different species. In *Foundations of Experimental Embryology*, eds., WILLIER, B. H. and OPPENHEIMER, J. M., pp. 145–84. Prentice-Hall, Englewood Cliffs, New Jersey.

245. SPEMANN, H. and SCHOTTÉ, O. (1932). Uber xenoplastische Transplanta-tion als Mittel zur Analyse der embryonalen Induction. *Naturwissen-schaft*, **20**, 463–67.

246. SPIRIN, A. S. (1966). On 'masked' focus of messenger RNA in early embryogenesis and other differentiation systems. *Current Topics in Developmental Biology*, eds. MOSCONA, A. A. and MONROY, A., Vol. 1, pp. 1–38. Academic Press, New York.

247. SRIVASTAVA, P. N., ADAMS, C. E. and HARTREE, E. F. (1965). Enzymatic action of lipoglycoprotein preparations from sperm acrosomes on rabbit ova. *Nature, Lond.* **205**, 498.

248. STONE, L. S. (1958). Regeneration of the retina, iris and lens. In *Regeneration in Vertebrates*, ed. THORNTON, C. S., pp. 3–14. University of Chicago Press, Chicago, Illinois.

249. STRUDEL, G. (1967). Some aspects of organogenesis of the chick spinal column. *Exp. Biol. Med.*, **1**, 183–98.

250. SUSSMAN, M. (1964). *Growth and Development*, 2nd ed. Prentice-Hall, Englewood Cliffs, New Jersey.

251. TIEDEMANN, H. (1966). The molecular basis of differentiation in early development of amphibian embryos. In *Current Topics in Developmental Biology*, eds. MOSCONA, A. A. and MONROY, A., Vol. 1, pp. 85–110. Academic Press, New York.

252. TIEDEMANN, H. (1967). Biochemical aspects of primary induction and determination. In *The Biochemistry of Animal Development*, ed. WEBER, R., Vol. 2, pp. 4–55. Academic Press, New York.

253. TIEDEMANN, H. (1967). Inducers and inhibitors of embryonic differen-tiation. Their chemical nature and mechanism of action. *Exp. Biol. Med.*, **1**, 8–21.

254. TOIVONEN, S. (1954). The inducing action of the bone marrow of the Guinea pig after alcohol and heat treatment in implantation and ex-plantation experiments with embryos of *Triturus*. *J. Embryol. exp. Morph.*, **2**, 239–44.

255. TOIVONEN, S. (1967). Mechanism of primary induction. *Exp. Biol. Med.*, **1**, 1–7.

256. TOIVONEN, S. and SAXEN, L. (1955). The simultaneous inducing action of liver and bone-marrow of the Guinea pig in implantation and explan-tation experiments with embryos of *Triturus*. *Expl Cell Res.* (suppl.), **3**, 346–57.

257. TOIVONEN, S. and SAXEN, L. (1957). Embryonic inductive action of normal and leukemic bone marrow of the rat. *J. natn. Cancer Inst.*, **19**, 1095–104.

258. TORREY, T. W. (1962). *Morphogenesis of the Vertebrates.* Wiley, New York.

259. TYLER, A. (1939). Extraction of an egg membrane lysin from sperm of the giant keyhole limpet (*Megathura crenulata*). *Proc. natn. Acad. Sci. U.S.A.*, **25**, 317–23.

260. TYLER, A. (1941). The role of fertilizin in the fertilization of eggs of the sea urchin and other animals. *Biol. Bull. mar. biol. Lab, Woods Hole*, **81**, 190–204.

261. TYLER, A. (1948). Fertilization and immunity. *Physiol. Rev.*, **28**, 180–219.

262. TYLER, A. (1949). Properties of fertilizin and related substances of eggs and sperm of marine animals. *Am. Nat.*, **83**, 195–219.

263. TYLER, A. (1957). Immunological studies of early development. In *The Beginning of Embryonic Development*, eds. TYLER, A., VON BORSTEL, R. C. and METZ, C. B. pp. 341–82. Am. Ass. Advance. Sci., Washington, D.C.

264. TYLER, A. (1960). Introductory remarks on theories of fertilization. In *Symposium on the Germ Cells and Earliest Stages of Development*, pp. 155–74. Instit. Intern. d'Embryologie and Fondazione A. Baselli, Instituto Lombardo, Milano, Italy.

265. TYLER, A. (1963). The manipulations of macromolecular substances during fertilization and early development of animal eggs. *Am. Zool.*, **3**, 109–26.

266. TYLER, A. (1965). The biology and chemistry of fertilization. *Am. Zool.*, **99**, 309–34.

267. TYLER, A. (1967). Masked messenger RNA and cytoplasmic DNA in relation to protein synthesis and process of fertilization and determination in embryonic development. *Devl Biol. Suppl. I*, 170–226.

268. TYLER, A. and BROOKBANK, J. W. (1956). Antisera that block cell division in developing eggs of sea urchins. *Proc. natn. Acad. Sci.*, **42**, 304–08.

269. VOGT, W. (1923). Weitere Versuche mit vitaler Farbmarkierung und farbiger Transplantation zur Analyse der Primitiventwicklung von Triton. *Verh. anat. Ges., Jena* (Anat. Anz. suppl.), **57**, 30–38.

270. VOGT, W. (1929). Gestaltungsanalyse am Amphibienkeim mit örtlicher Vitalfärbung. II. Gastrulation und Mesodermbildung bei Urodelen und Anuren. *Arch. EntwMech. Org.*, **120**, 384–706.

271. WADDINGTON, C. H. (1947). *Organizers and Genes.* Cambridge University Press, London.

272. WADDINGTON, C. H. (1956). *Principles of Embryology.* Macmillan, New York.

273. WADDINGTON, C. H. (1962). *New Patterns in Genetics and Development.* Columbia University Press, New York.

274. WADDINGTON, C. H. (1966). *Principles of Development and Differentiation.* Macmillan, New York.

275. WADDINGTON, C. H. and SIRLIN, J. L. (1954). The incorporation of labelled amino acids into Amphibian embryos. *J. Embryol. exp. Morph.*, **2**, 340–47.

276. WADDINGTON, C. H. and SIRLIN, J. L. (1955). Studies in Amphibian embryogenesis using labelled grafts. *Proc. R. phys. Soc. Edinb.*, **24**, 28–31.

277. WARBURG, O. (1910). Über die Oxydationen in lebenden Zellen nach Versuchen am Seeigelei. *Hoppe-Seyler's Z. physiol. Chem.*, **66**, 305.

278. WEBER, R. (1965). *The Biochemistry of Animal Development.* Academic Press, New York.

279. WEISS, P. A. (1939). *Principles of Development.* Holt, Rinehart and Winston, New York.

280. WEISS, P. (1949). Nature of vertebrate individuality. The problem of cellular differentiation. *Proc. natn. Cancer Conf.*, pp. 50–60.

281. WEISS, P. (1950). Perspectives in the field of morphogenesis. *Q. Rev. Biol.*, **25**, 177–98.

282. WEISS, P. (1953). Some introductory remarks on the cellular basis of differentiation. *J. Embryol. exp. Morph.*, **1**, 181–211.

283. WEISS, P. (1959). Discussion. *In Biological Organization, Cellular and Subcellular.* Proc. Symp. at the Univ. Edinburgh, 1957, ed. WADDINGTON, C. H., p. 231. Pergamon Press, London.

284. WEISS, P. (1968). Dynamics of Development, Experiments and Inferences. Academic Press, New York.

285. WESSELLS, N. K. (1964). DNA synthesis, mitosis, and differentiation in pancreatic acinar cells *in vitro. J. cell Biol.*, **20**, 415–33.

286. WESSELLS, N. K. (1964). Substrate and nutrient effects upon epidermal basal cell orientation and proliferation. *Proc. natn. Acad. Sci. U.S.A.*, **52**, 252–59.

287. WESSELLS, N. K. (1964). Tissue interactions and cytodifferentiation. *J. exp. Zool.*, **157**, 139–52.

288. WESSELLS, N. K. (1968). Problems in the analysis of determination, mitosis and differentiation. In *Epithelial–Mesenchymal Interactions*, eds. FLEISCHMAJER, R. and BILLINGHAM, R., pp. 132–50. The Williams and Wilkins Co., Baltimore.

289. WESSELLS, N. K. and COHEN, J. H. (1967). Early pancreas organogenesis: morphogenesis, tissue interactions and mass effects. *Devl Biol.*, **15**, 237–70.

290. WESSELLS, N. K. and WILT, F. H. (1965). Action of actinomycin D on exocrine pancreas cell differentiation. *J. molec. Biol.*, **13**, 767–79.

291. WHITAKER, D. M. (1931). On the rate of oxygen consumption by fertilized and unfertilized eggs. I. *Fucus vesiculosus. J. gen. Physiol.*, **15**, 167–82.

292. WHITAKER, D. M. (1933). On the rate of oxygen consumption by fertilized and unfertilized eggs. V. Comparison and interpretation. *J. gen. Physiol.*, **16**, 497–528.

293. WILLIER, B. H. and OPPENHEIMER, J. M. (1964). *Foundations of Experimental Embryology.* Prentice-Hall, Englewood Cliffs, New Jersey.

294. WILLIER, B., WEISS, P. and HAMBURGER, V. (1955). *Analysis of Development.* W. B. Saunders, Philadelphia.

295. WILT, F. H. (1966). The concept of messenger RNA and cytodifferentiation. *Am Zool.*, **6**, 67–74.

296. WILT, F. H. and HULTIN, T. (1962). Stimulation of phenylalanine incorporation by polyuridylic acid in homegenates of sea urchin eggs. *Biochem. biophys. Res. Commun.*, **9**, 313–17.

297. WILT, F. and WESSELLS, N. K. (1967). *Methods in Developmental Biology*. Thomas Y. Crowell Co., New York.

298. WITTMAN, K. S. and HAMBURGH, M. (1968). The development and effect of genetic background on expressivity and penetrance of the brachyury mutation in the mouse: A study in developmental genetics. *J. exp. Zool.*, **168**, 137–45.

299. WOLFF, E. (1966). General factors of embryonic induction. In *Cell Differentiation and Morphogenesis*. Int. Lecture Course, Wageningen, The Netherlands, pp. 1–19. North Holland Publishing Co., Amsterdam.

300. WOLFF, E. (1968). Tissue interactions during organogenesis. In *Current Topics in Developmental Biology*, eds. MOSCONA, A. A. and MONROY, A., Vol. 3, pp. 65–93. Academic Press, New York.

301. YAMADA, T. (1958). Induction of specific differentiation by samples of proteins and nucleoproteins in the isolated ectoderm of *Triturus* gastrulae. *Experientia*, **14**, 81–87.

302. YAMADA, T. (1958). Embryonic Induction. In *A Symposium on Chemical Basis of Development*, eds. MCELROY, W. and GLASS, B., pp. 217–38. Johns Hopkins Press, Baltimore.

303. YAMADA, T. (1962). The inductive phenomenon as a tool for understanding the basic mechanisms of differentiation. *J. Cell. comp. Physiol.* Suppl. 1, **60**, 49–64.

304. YAMADA, T. (1967). Cellular and subcellular events in Wolffian lens regeneration. In *Current Topics in Developmental Biology*, eds. MOSCONA, A. A. and MONROY, A., Vol. 2, pp. 247–81. Academic Press, New York.

305. YAMADA, T. (1967). Cellular synthetic activities in induction of tissue transformation. In *Cell Differentiation, CIBA Fdn Symp.*, eds. DE REUCK, A. V. S. and KRUGEL, J., pp. 116–30. Little, Brown and Company, Boston.

306. YAMADA, T. and TAKATA, K. (1955). Effect of trypsin and chymotrypsin on the inducing ability of the kidney and its fractions. *Expl Cell Res.*, **3**, 402–13.

307. ZILLIKEN, F. (1967). Notochord induced cartilage formation in chick somites. Intact tissue versus extracts. *Exp. Biol. Med.*, **1**, 199–212.

308. ZWILLING, E. (1942). The development of dominant Rumplessness in chick embryos. *Genetics, Princeton*, **27**, 641–56.

309. ZWILLING, E. (1945). The embryogeny of a recessive rumpless condition of chickens. *J. exp. Zool.*, **99**, 79–91.

310. ZWILLING, E. (1956). Genetic mechanism in limb development. *Cold Spring Harb. Symp. quant. Biol.*, **1**, 349–54.

311. ZWILLING, E. (1956). Interaction between limb bud ectoderm and mesoderm in the chick embryo. II. Experimental limb duplication. *J. exp. Zool.*, **132**, 173–87.

312. ZWILLING, E. (1956). Interaction between limb bud ectoderm and mesoderm in the chick embryo. IV. Experiments with a wingless mutant. *J. exp. Zool.*, **132**, 241–53.

313. ZWILLING, E. (1961). Limb morphogenesis. In *Advances in Morphogenesis*, eds. ABERCROMBIE, M. and BRACHET, J., Vol. 1, pp. 301–30. Academic Press, New York.

314. ZWILLING, E. and HANSBOROUGH, L. (1956). Interaction between limb bud ectoderm and mesoderm in the chick embryo. III. Experiments with polydactylous limbs. *J. exp. Zool.*, **132**, 219–39.

Index

Index

Major entries and pages on which definitions are found are shown in **bold** type, those on which illustrations are found are shown in *italics*.